Thyretini of Africa

Łukasz Przybyłowicz

Thyretini of Africa

An Illustrated Catalogue of the Thyretini
(Lepidoptera: Arctiidae: Syntominae)
of the Afrotropical Region

Entomonograph Volume 16
A series facing global biodiversity in insects

Apollo Books
Denmark, 2009

Printed by Vinderup Bogtrykkeri ApS, Vinderup

Bound by J. P. Møller Bogbinderi A/S, Haderslev

Published by:
Apollo Books
Kirkeby Sand 19
DK-5771 Stenstrup
Denmark
Telephone 0045 6226 3737
Fax 0045 6226 3780
apollobooks@vip.cybercity.dk
www.apollobooks.com

ISBN 978-87-88757-91-0
ISSN 0106-2808

Author's address:
Dr. Łukasz Przybyłowicz
Polish Academy of Sciences
Institute of Systematics and Evolution of Animals
Slawkowska 17
PL-31-016 Cracow
Poland
lukasz@isez.pan.krakow.pl

TABLE OF CONTENTS

1. Acknowledgements 6

2. Summary 7

3. Introduction 7

4. Arrangement of the book 7

5. List of abbreviations of institutions 9

6. Catalogue 10

6. Unavailable names 30

7. Taxa excluded from the Thyretini 32

8. Taxonomic changes and comments to the generic level categories 34

9. Descriptions of species 37

10. References 106

11. Colour plates with imago habitus 114

12. Plates with male genitalia 126

13. Plates with female genitalia 144

14. Index of scientific names in Lepidoptera 163

Acknowledgements

The work published in this catalogue was supported by grant nr 2 P04C 034 26 of the Committee for Scientific Research (KBN), Warsaw, Poland.

I cordially thank the lepidopterist colleagues who have provided valuable information and materials for this study (alphabetically): Daniel Burckhardt (Basel), Ugo Dall'Asta (Tervuren), Rienk De Jong (Leiden,) Jurate de Prins (Tervuren,) Barbara Dombrowsky (Pretoria), David Goodger (London), Bert Gustafsson (Stockholm), Joachim Haendel (Halle), Don Harvey (Washington), Axel Hausmann (Munich), Martin Honey (London), Ole Karsholt (Copenhagen), Martin Krüger (Pretoria), Bernard Landry (Geneva), Paul Limbourg (Brussels), Martin Lödl (Wien), Darren J. Mann (Oxford), Geoff Martin (London), Wolfram Mey (Berlin), Joël Minet (Paris), Violah Muyambo (Bulawayo), Roberto Poggi (Genoa), Suzanne Rab Green (New York), John Rawlins (Pittsburgh), Hans Riefenstahl (Hamburg), Andreas Segerer (Munich) and Sergey Sinev (St. Petersburg).

I also express my thanks to my friend Marek Kopeć (Kraków) for his kind help in preparing the photographs of some genitalia and Krzysztof Fiołek (Kraków) for his great help in arranging all tables.

My special thanks go to Bernard Landry (Geneva) for his editorial effort and for linguistic corrections on the manuscript.

I acknowledge very much the kindness, hospitality and very nice atmosphere created by my friends and the remaining workers of the RMCA during my several long and short visits to Tervuren. Ugo Dall'Asta was always willing to help me with any problem, Frans Desmet was the source of knowledge of all strange localities, and Jurate and Willy de Prins were very nice companions and brought "new spirit" in my taxonomical work.

I also express my gratitude to Martin Honey (London) who always answered my numerous mails in details, explaining many difficult problems.

SUMMARY

With the present book a comprehensive, illustrated catalogue of the tribe Thyretini (Lepidoptera: Arctiidae: Syntominae) is presented. In the first part each of the 194 recognized species are listed according to their current generic combination. An additional 138 synonymic species-group names are mentioned (in total 332 species-group names). The species are grouped in 21 genera, for which 19 synonyms are additionally mentioned. Information on the original description, type locality and type specimens is given. 7 genus-group names and 62 species-group names are synonymized.

INTRODUCTION

The tribe Thyretini is a small group of middle to large-sized arctiids, restricted to the African continent, with only two species reaching the Arabian Peninsula.

The name Thyretinae was proposed by BUTLER (1876) for the subfamily established in Zygaenidae (see also KIRBY 1892). However, HAMPSON (1898) transfered the genus *Thyretes* BOISDUVAL to the Syntomidae [Ctenuchidae]. It remained there until KIRIAKOFF (1949) created the family Thyretidae for *Thyretes* and other syntomines in which the tympanum faces ventrally, as in Notodontidae. KIRIAKOFF (1960b) listed 25 genera including 203 species, all of them African. Based on similarities of the tympanum, KIRIAKOFF associated thyretids with notodontids despite numerous other differences between them. With reservation, MINET (1986) formally placed thyretines as a subfamily of the Notodontidae.

Based on modified pheromone glands in the females, a tymbal organ, and arctiid-like larvae, HOLLOWAY (1988) returned Thyretinae to Arctiidae considered them to be the probable sister-group of Syntominae [Syntomini]. KITCHING & RAWLINS (1998) treated the Thyretinae as a tribe of the Syntominae. The same results were obtained by JACOBSON & WELLER (2002) basing on the phylogenetic study of the world Arctiidae.

Nowadays, according to the broad molecular and morphological studies Arctiidae are more and more frequently considered to be a subfamily of the enlarged, monophyletic family Noctuidae. In this system Thyretini is downgraded to the subtribe Thyretina (LAFONTAINE & FIBIGER 2006). However, in the present catalogue the traditional classification is still applied.

The most productive author working on Thyretines was KIRIAKOFF who proposed 32 genus-group categories and described 148 new species and subspecies. However his type series were very often based on material too limited for him to appreciate that many of his supposed species were within the normal range of specific variation of a single species. Furthermore, in several cases he did not recognize the sex of described specimens correctly and some specimens had not been labeled by him as types. All this causes that although the group is relatively small, more than 25% of species are of doubtful systematic position. Systematic revisions of genera *Metarctia* and *Rhipidarctia* are highly required.

In the present catalogue several new synonyms and new combinations are introduced. All of them are based on examination of types.

ARRANGEMENT OF THE BOOK

I. Catalogue

In the present catalogue, the genus names are listed in alphabetical order throughout. If, in some cases these are further divided into subgenera, they are listed in the same way. For each genus (subgenus), the original spelling was checked, the reference to the original description is given, as is the type species and the manner of its designation. Also, all known synonyms are listed chronologically with the same data.

Within each genus (subgenus) the species are listed alphabetically. The original description was checked for all listed names of the species group and as in many modern catalogues (e.g. SCOBLE 1999, DE PRINS & DE PRINS 2005), the original spelling was retained, regardless of the gender of the current generic combination. The species entries include author, year of publication, original genus in parentheses, type designation (e.g. syntype, lectotype, holotype), country (according to the present political division) of type locality, type locality in original spelling (as on the label) and institution of deposition of the type. Following each species, its synonyms are listed chronologically. Subordinate taxa recognized as valid subspecies are indicated in bold.

All valid species names are provided with Arabic numerals starting from 1, which are followed in the next parts of the book. Therefore every species bears its own, always the same number of description, imago and genitalia illustration.

II. Descriptions of species.

The second part of the book provides short information about each species. Species are arranged as in the catalogue and are provided with the same numbers.

Short diagnosis. Imago. The most characteristic features separating the taxon from the others are listed. In some cases available material does not allow for unquestionable interpretation of very similar taxa. They are however not synonymized as such action should be preceded by revision based on more extensive material. In some cases different characters for male and female are provided.

Male genitalia. Only the most important morphological characters are pointed out. In several cases when the taxon is known from a unique specimen and is easily separated from the others, the genitalia were not dissected.

Female genitalia. As for the males. In many cases the females remain unknown so the diagnostic characters listed for these known, should be treated with reserve.

Early stages. Short information about published data concerning developmental stages are provided. In all cases the references are included.

Biology. All available information on host plants, life cycle, flying period and daily activity are provided. They are mostly retrieved from the literature but in some cases they relay on personal observations of collectors.

Distribution. The distribution is alphabetically listed by names of countries which in some cases are shortened e.g. DRC (Democratic Republic of the Congo), RSA (Republic of South Africa).

Remarks. Here the explanations for some of the taxonomic decisions as well as all other information which might be useful for better understanding of the given taxon are provided.

III. References

The bibliography includes all published references that comprise original descriptions of taxa. It includes also all references known to treat various aspects of the nomenclature, taxonomy, faunistics, biology or any other information concerning Thyretini, whether or not this information is expressly cited in the catalogue. All references cited in the catalogue proper were examined, either as originals or photocopies.

IV. Plates

Colour plates depicting the imagos are followed by plates with male and female genitalia. First of these parts is numbered with Romen numerals. Remaining two parts are numbered continuously with Arabic numerals. Each species bears the same number as in the catalogue and descriptions.

When possible the type specimens (holotype, paratype, lectotype, paralectotype or syntype) are illustrated rather than other specimens. This option was chosen to prevent the misidentification of numerous similar taxa of controversial systematic position described mostly by KIRIAKOFF.

LIST OF ABBREVIATIONS OF INSTITUTIONS

AMNH – American Museum of Natural History, New York, USA

BMNH – The Natural History Museum, London, U.K.

BMZ – The Natural History Museum, Bulawayo, Zimbabwe

CMNH – Carnegie Museum of Natural History, Pittsburgh, USA

ISEA – Institute of Systematics and Evolution of Animals, Cracow, Poland

KBIN – Royal Belgian Institure of Natural Sciences, Brussels, Belgium

MCSNG – Museo Civico di Storia Naturale, Genoa, Italy

MHNG – Muséum d'Histoire naturelle, Genève, Switzerland

MNHN – Muséum national d'Histoire naturelle, Paris, France

MLU – Martin Luther University, Halle, Germany

MRSN – Museo Regionale di Scienze Naturali, Torino, Italy

NHMB – Naturhistorisches Museum, Basel, Switzerland

NHMW - Naturhistorisches Museum, Wien, Austria

NHRS – Naturhistoriska Riksmuseet, Stockholm, Sweden

OUM – Hope Museum, Oxford University Museum of Natural History, U.K.

RMCA – Royal Museum for Central Africa, Tervuren, Belgium

RNHL - National Natural History Museum Naturalis, Leiden, The Netherlands

SMNS – Staatliches Museum für Naturkinde, Stuttgart, Germany

TMSA – Transvaal Museum, Pretoria, South Africa

UGD – Ernst Moritz Arndt University of Greifswald, Germany

USNM - National Museum of Natural History, Washington D. C., USA

ZIMH – Zoologisches Institut und Museum, Hamburg, Germany

ZIN – Zoological Institute of the Russian Academy of Sciences, St. Petersburg, Russia

ZMHB – Zoologisches Museum der Humboldt Universität, Berlin, Germany

ZMUC – Zoological Museum, University of Copenhagen, Denmark

ZSM – Zoologische Sammlung des Bayerischen Staates, Munich, Germany

Family: Arctiidae
Subfamily: Syntominae
Tribe: *Thyretini*

Apisa WALKER, 1855b: 916-917.
Type species: *Apisa canescens* Walker, 1855 (by monotypy).

Subgenus *Apisa* WALKER, 1855b: 916-917.
Type species: *Apisa canescens* Walker, 1855 (by monotypy).

1. *arabica* WARNECKE, 1934: 63 (*Apisa*). Syntypes [Yemen] „San'aa" [ZIMH].
> *lippensi* KIRIAKOFF, 1960a: 4 (*Apisa canescens* ssp.). Holotype [Saudi Arabia] "Habne (ville)" [KBIN – not located, Ł. Przybyłowicz]. **[syn. nov.]**

2. *canescens* WALKER, 1855b: 917 (*Apisa*). Lectotype **here designated** "S. Africa" [BMNH].
> *pallata* PLÖTZ, 1880: 78 (*Psychotoe*). Holotype [Congo] „W.-Afrika, Abo" [UGD] (synonymized by Hampson 1898: 142).
> *cana* HOLLAND, 1893: 394-395 (*Apisa*). Syntypes [Gabon] "Kangwé, Ogové Riv." [CMNH] (synonymized by Hampson 1898: 142).
> *microcanescens* BERIO, 1935: 59 (*Apisa canescens* ssp.). Holotype [Somalia] " Belet Amin. Giuba, " [MCSNG].
> *tamsi* KIRIAKOFF, 1957b: 122 (*Apisa*). Holotype [Burundi] „Kasenyi" [RMCA]. **[syn. nov.]**

Subgenus *Dufraneella* KIRIAKOFF, 1953b: 14-15.
Type species: *Metarctia grisescens* DUFRANE, 1945 (by original designation).

3. *fontainei* KIRIAKOFF, 1959b: 25 (*Apisa*). Holotype [Rwanda] „Kisenyi" [RMCA].

4. *grisescens* DUFRANE, 1945: 131 (*Metarctia*). Holotype [DRC] „Kamituga" [KBIN].

5. *hildae* KIRIAKOFF, 1961b: 96-97 (*Apisa*). Holotype [Namibia] „Okahandja" [ZSM]. **[comb. nov.]**

6. *rendalli* ROTHSCHILD, 1910: 441-442 (*Apisa*). Lectotype **here designated** [Malawi] "Zomba, Upper Shire River" [BMNH]. **[comb. nov.]**
> *nyasae* KIRIAKOFF, 1957c: 95 (*Apisa grisescens* ssp.). Holotype [Malawi] "Nyasaland, near Mlanje" [BMNH]. **[syn. nov.]**

7. *subcanescens* ROTHSCHILD, 1910: 442 (*Apisa*). Lectotype **here designated** [Senegal] "Casamance" [BMNH].

Subgenus *Parapisa* KIRIAKOFF, 1952d: 174.
Type species: *Parapisa bourgognei* KIRIAKOFF, 1952d (by original designation).

8. *cinereocostata* HOLLAND, 1893: 394 (*Apisa*). Holotype [Gabon] "Valley of the Ogove River" [CMNH]. **[stat. rev.] [comb. nov.]**

        bourgognei KIRIAKOFF, 1952d: 173-174 (*Apisa*). Holotype [Ivory Coast] "Bingerville" [MNHN]. **[syn. nov.]**

9. *subargentea* JOICEY & TALBOT, 1921: 158 (*Apisa*). Holotype [Rwanda] "Lake Tshohoa" [BMNH].

Apisa INCERTAE SEDIS

10. *manettii* TURATI, 1924: 46-49 (*Apisa*). Syntypes [Libya] "Cyrenaica: Bengasi-Fuehat, giardino Vella ad El Berca" [MRSN].

Automolis HÜBNER, [1819] (1816-[1826]): 170.
Type species: *Sphinx meteus* STOLL, 1781 (by subsequent designation by KIRBY 1892: 220).
        Zagaris WALKER, 1855: 1096.
        Type species: *Zagaris crassa* FELDER, 1868 (by monotypy).
        Decimia WALKER, 1856: 1717-1718.
        Type species: *Decimia bicolora* WALKER, 1856 (by monotypy).

11. *bicolora* WALKER, 1856: 1718 (*Decimia*). Syntypes [RSA] "South Africa; Port Natal" [BMNH].

12. *crassa* FELDER, 1874: 10; pl. 99, fig. 16 (*Zagaris*). Holotype [RSA] "Knysna" [BMNH].
        meteus WALKER, 1855 [nec STOLL, 1780-82]: 1096 (*Zagaris*). Syntypes [RSA] "Cape" [BMNH] (synonymized by Zerny 1912a: 42).

13. *incensa* WALKER, 1864: 316 (*Anace*). Holotype [no locality] [BMNH].
        hewitti JANSE, 1945: 97-98 (*Metarctia*). Holotype [RSA] "Rhyn's Dorp (nearly 200 miles North of Cape Town)" [TMSA] (synonymized by Kiriakoff 1957b: 132).

14. *meteus* STOLL, 1780-82: 109-110 (*Sphinx*). Syntype(s) [RSA] "Kaap de Goede Hoop" [BMNH - not located (pers. comm. from M. Honey)].

15. *pallida* HAMPSON, 1901: 170 (*Metarctia*). Syntypes [Kenya] "Kikuyu, Nairobi, Romoro" [BMNH]. **[comb. nov.]**
        subrosea KIRIAKOFF, 1957c: 105 (*Metarctia*). Holotype [Kenya] "Njoro" [BMNH]. **[syn. nov.] [comb. nov.]**

Balacra WALKER, 1856: 1721.
Type species: *Balacra caeruleifascia* WALKER, 1856 (by subsequent designation by KIRBY 1892: 221).

Subgenus *Balacra* WALKER, 1856: 1721.
Type species: *Balacra caeruleifascia* WALKER, 1856 (by subsequent designation by KIRBY 1892: 221).

Megapisa Aurivillius, 1904: 29.

Type species: *Megapisa nigripennis* Aurivillius, 1904 (by monotypy).

Metapiconoma Rothschild, 1910: 444.

Type species: *Metapiconoma rattrayi* Rothschild, 1910 (by subsequent designation by Watson et al. 1980: 115).

16. *belga* Kiriakoff, 1954a: 3-4 (*Balacra*). Holotype [DRC] "Lupweji" [KBIN].

17. *caeruleifascia* Walker, 1856: 1721-1722 (*Balacra*). Holotype [no locality] [BMNH].

> *ochracea* Walker, 1869: 331 (*Balacra*). Holotype "Congo" [BMNH - not located (pers. comm. from M. Honey)]. **[syn. nov.]**
>
> *ehrmanni* Holland, 1893: 535-536 (*Automolis*). Holotype „Liberia" [CMNH]. **[syn. nov.]**
>
> *conradti* Oberthür, 1911: 471 (*Pseudapiconoma*). Syntypes [Cameroon] „Lolodorf" [BMNH] (synonymized with *B. ochracea* by Hampson 1914a: 80).
>
> *germana* Rothschild, 1912a: 119 (*Balacra*). Holotype "Sierra Leone" [BMNH] (synonymized with *B. ochracea* by Kiriakoff 1957b: 149).
>
> *inflammata* Hampson, 1914a: 80 (*Balacra*). Holotype [Ghana] "Aburi" [BMNH]. **[syn. nov.]**
>
> *magna* Hulstaert, 1923: 407 (*Balacra*). Holotype [DRC] "Beni" [RMCA] (synonymized with *B. ochracea* by Kiriakoff 1953b: 60).
>
> *similis* Hulstaert, 1923: 408 (*Balacra*). Holotype [DRC] "Leopoldville" [RMCA] (synonymized with *B. ochracea* by Kiriakoff 1953b: 60).

18. *guillemei* Oberthür, 1911: 469 (*Pseudapiconoma*). Holotype [DRC] "Tanganyika, M'pala" [BMNH]. **[stat. rev.]**

> *erubescens* Joicey & Talbot, 1924: 549-550 (*Balacra*). Syntypes [DRC] "Luvua River, 85 miles north of Lake Mweru" [BMNH]. **[syn. nov.]**

19. *nigripennis* Aurivillius, 1904: 30 (*Megapisa*). Holotype "Centralafrika" [NHRS].

> *gloriosa* Jordan, 1904: 441 (*Pseudapiconoma*). Holotype [Angola] "Pungo Andongo" [BMNH] (synonymized by Hampson 1914a: 73).
>
> *aurivilliusi* Kiriakoff, 1957b: 151-152 (*Balacra*). Holotype [DRC] "Tanganika: Kongolo" [RMCA]. **[syn. nov.]**

20. *rattrayi* Rothschild, 1910: 444 (*Metapiconoma*). Syntypes [Uganda] "Entebbe; Kurunga, Kyanika, Bulamwezi; Weni, River Toru" [BMNH].

Subgenus *Callobalacra* Kiriakoff, 1953b: 70.

Type species: *Metarctia rubrostriata* Aurivillius, 1898 (by original designation).

21. *alberici* Dufrane, 1945: 134 (*Balacra*). Holotype [DRC] „Kamituga" [KBIN]. **[stat. rev.]**

22. *jaensis* Bethune-Baker, 1927: 322 (*Balacra*). Holotype [Cameroon] "Bitje" [BMNH].

23. *rubrostriata* Aurivillius, 1898: 185-186 (*Metarctia*). Holotype "Togo" [NHRS].

Subgenus *Compsochromia* KIRIAKOFF, 1953b: 71.

Type species: *Pseudapiconoma compsa* JORDAN, 1904 (by original designation).

24. *compsa* JORDAN, 1904: 441 (*Pseudapiconoma*). Holotype [Angola] "Pungo Andongo" [BMNH].

    fenestrata JORDAN, 1904: 442 (*Pseudapiconoma*). Holotype [Angola] "Pungo Andongo" [BMNH] (synonymized by Kiriakoff 1957b: 153).

    melaena HAMPSON, 1905: 426 (*Pseudapiconoma*). Holotype [Uganda] "Masaka" (BMNH) (synonymized with *B. fenestrata* by Seitz 1926: 56).

    stigmatica GRÜNBERG, 1907: 435-436 (*Pseudapiconoma*). Holotype „Uganda" [ZMHB] (synonymized by Seitz 1926: 56).

    vitreata ROTHSCHILD, 1910: 445 (*Pseudapiconoma*). Holotype [locality unknown] [BMNH] (synonymized by Kiriakoff 1957b: 153).

25. *diaphana* KIRIAKOFF, 1957c: 112 (*Balacra*). Holotype [Uganda] "Kawanda" [BMNH].

Subgenus *Daphaenisca* KIRIAKOFF, 1953b: 69.

Type species: *Pseudapiconoma daphaena* HAMPSON, 1898 (by original designation).

    Balacrella KIRIAKOFF, 1957b: 154. **[syn. nov.]**

    Type species: *Pseudapiconoma affinis* ROTHSCHILD, 1910 (by original designation).

26. *affinis* ROTHSCHILD, 1910: 442 (*Pseudapiconoma*). Holotype [DRC] "Kassai District, Congo Free State" [BMNH]. [comb. nov.]

27. *daphaena* HAMPSON, 1898: 150 (*Pseudapiconoma*). Holotype [Nigeria] "R. Niger, Asaba" [BMNH].

Subgenus *Heronina* KIRIAKOFF, 1955b: 265.

Type species: *Anace herona* DRUCE, 1887 (by original designation).

28. *herona* DRUCE, 1887: 669 (*Anace*). Syntype(s) [Cameroon] "Mongo-ma Lubah", [Ghana] "Aburi" [BMNH].

Subgenus *Lamprobalacra* KIRIAKOFF, 1953b: 72.

Type species: *Balacra pulchra* AURIVILLIUS, 1892 (by original designation).

29. *elegans* AURIVILLIUS, 1892: 190 (*Balacra*). Holotype "Camerun" [NHRS]

    damalis HOLLAND, 1893: 397 (*Balacra*). Syntypes [Equatorial Guinea, Zambia] "Benita and Kangwe" [CMNH] (synonymized by Hampson 1898: 152).

    curriei DYAR, 1899: 174 (*Pseudapiconoma elegans* var.). Syntypes [Liberia] "Mt. Coffee" [USNM] (synonymized by Kiriakoff 1953b: 73-74).

    bumba STRAND, 1918: 6 (*Pseudapiconoma*). Holotype [DRC] "Bumba" [ZMHB]. **[syn. nov.]**

30. *furva* HAMPSON, 1911: 394 (*Balacra*). Holotype [Ghana] "Gold Coast" [BMNH].

31. *pulchra* AURIVILLIUS, 1892: 200 (*Balacra*). Holotype "Camerun" [NHRS]

  glagoessa HOLLAND, 1893: 396 (*Balacra*). Syntype(s) [Gabon] "Valley of the Ogove River" [CMNH] (synonymized by Hampson 1914a: 82).

32. *rubricincta* HOLLAND, 1893: 396 (*Balacra*). Holotype [Gabon] "Valley of the Ogove River" [CMNH].

Subgenus *Pseudapiconoma* AURIVILLIUS, 1881: 46.
Type species: *Pseudapiconoma testacea* AURIVILLIUS, 1881 (by monotypy).

  Epibalacra KIRIAKOFF, 1957b: 149. **[syn. nov.]**
  Type species: *Metarctia preussi* AURIVILLIUS, 1904 (by original designation).

33. *basilewskyi* KIRIAKOFF, 1953b: 59 (*Balacra*). Holotype [DRC] "Bena-Dibele" [RMCA].

34. *batesi* DRUCE, 1910: 393 (*Pseudapiconoma*). Holotype [Cameroon] "Bitje, Ja River" [BMNH].

  congoensis ROTHSCHILD, 1910: 443 (*Pseudapiconoma batesi* ssp.). Holotype [DRC] "Luebo, Kassai" [BMNH] (synonymized by Zerny 1912a: 45).

  ugandae ROTHSCHILD, 1910: 443 (*Pseudapiconoma batesi* ssp.). Holotype [Uganda] "Entebbe" [BMNH] (synonymized by Zerny 1912a: 45).

  decora OBERTHÜR, 1911: 470 (*Pseudapiconoma flavimacula* var.). Syntypes [Cameroon] "Johann-Albrechts Höhe" [BMNH]. **[syn. nov.]**

  distincta KIRIAKOFF, 1953b: 66 (*Balacra*). Holotype [DRC] "Lusambo" [RMCA]. **[syn. nov.]**

35. *flavimacula* WALKER, 1856: 1722 (*Balacra*). Holotype [Ghana] "Ashanti" [BMNH].

  testacea AURIVILLIUS, 1881: 46 (*Pseudapiconoma*). Holotype [Gabon] "Gaboon" [NHRS] (synonymized by Przybyłowicz & Kühne 2008: 148).

36. *fontainei* KIRIAKOFF, 1953b: 66-67 (*Balacra*). Holotype [DRC] "Lusambo" [RMCA].

  oreophila KIRIAKOFF, 1963a: 97 (*Balacra fontainei* ssp.). Holotype [DRC] "piste Ruwenzori" [RMCA]. **[syn. nov.]**

37. *haemalea* HOLLAND, 1893: 397 (*Balacra*). Holotype [Gabon] "Valley of the Ogove River" [CMNH].

38. *humphreyi* ROTHSCHILD, 1912a: 119-120 (*Balacra*). Holotype [Nigeria] "Ilesha" [BMNH].

  intermedia ROTHSCHILD, 1912a: 120 (*Balacra*). Holotype "Sierra Leone" [BMNH]. **[syn. nov.]**

39. *monotonia* STRAND, 1912: 191-192 (*Pseudapiconoma flavimacula* var.). Syntypes [Equatorial Guinea] "Alen; Makomo, Ntumgebiet" [ZMHB]. **[stat. res.]**

  simplex AURIVILLIUS, 1925b: 1303 (*Balacra*). Holotype [Congo] "Ouesso" [ZIMH – destroyed during the Second World War (pers. comm. from H. Riefenstahl)]. **[syn. nov.]**

  simplicior KIRIAKOFF, 1957a: 280 (*Balacra*). Holotype [DRC] "Sankuru: Luluabourg" [RMCA]. **[syn. nov.]**

  angolensis KIRIAKOFF, 1961b: 108-109 (*Balacra rubrovitta* ssp.). Holotype [Angola] "Bolongongo"

[ZSM]. **[syn. nov.]**

40. *preussi* AURIVILLIUS, 1904: 31 (*Metarctia*). Holotype [Cameroon] "Buea" [NHRS].

speculigera GRÜNBERG, 1907: 434 (*Pseudapiconoma*). Holotype [Cameroon] "Jaunde-Station" [ZMHB] (synonymized with *P. umbra* by Seitz 1926: 55).

laureola DRUCE, 1910: 393-394 (*Pseudapiconoma*). Syntypes [Cameroon] „Bitje, Ja River" [BMNH]. **[syn. nov.]**

umbra DRUCE, 1910: 394 (*Pseudapiconoma*). Holotype [Cameroon] „Bitje, Ja River" [BMNH] (synonymized by Kiriakoff 1957b: 150).

vitreigutta HULSTAERT, 1923: 407-408 (*Balacra*). Holotype [DRC] „Leopoldville" [RMCA] (synonymized by Kiriakoff 1953b: 62).

Bergeria KIRIAKOFF, **1952b: 397.**

Type species: *Bergeria haematochrysia* KIRIAKOFF, 1952 (by original designation).

41. *bourgognei* KIRIAKOFF, 1952b: 401-402 (*Bergeria*). Holotype [DRC] „Basoko" [RMCA].

fletcheri KIRIAKOFF, 1957c: 113 (*Bergeria*). Holotype [DRC] „Lisala" [BMNH]. **[syn. nov.]**

42. *haematochrysia* KIRIAKOFF, 1952b: 397-399 (*Bergeria*). Holotype [DRC] „Sankuru: Lusambo" [RMCA].

occidentalis KIRIAKOFF, 1957b: 156 (*Bergeria haematochrysia* ssp.). Holotype [Cameroon] "Efulen" [BMNH].

43. *octava* KIRIAKOFF, 1961a: 11-12 (*Bergeria*). Holotype [DRC] "Ubangi, region de Bumba" [RMCA].

44. *ornata* KIRIAKOFF, 1959b: 33 (*Bergeria*). Holotype [DRC] „Uele: Paulis" [RMCA].

45. *schoutedeni* KIRIAKOFF, 1952b: 401 (*Bergeria*). Holotype [DRC] "Eala" [RMCA].

46. *tamsi* KIRIAKOFF, 1952b: 399-400 (*Bergeria*). Holotype [DRC] „Equateur: Mondombe" [RMCA].

Cameroonia PRZYBYŁOWICZ **[gen. nov.]**

Type species: *Metarctia nigriceps* AURIVILLIUS, 1904 (by present designation)

47. *nigriceps* AURIVILLIUS, 1904: 30 (*Metarctia*). Holotype [Cameroon] "Camerun" [NHRS].

Hippurarctia KIRIAKOFF, **1953b: 32.**

Type species: *Hippurarctia vicina* KIRIAKOFF, 1953b (by original designation).

48. *cinereoguttata* STRAND, 1912: 189 (*Metarctia*). Holotype [Equatorial Guinea] "Nkolentangan" [ZMHB].

15

cameruna HAMPSON, 1914a: 65 (*Metarctia*). Syntypes "Cameroons" [BMNH] (synonymized by Kiriakoff 1959b: 28).

49. *ferrigera* DRUCE, 1910: 395 (*Metarctia*). Holotype [Cameroon] "Bitje, Ja River" [BMNH].

    kamitugensis DUFRANE, 1945: 133 (*Metarctia taymansi* f. n.). Holotype [DRC] "Kamituga" [KBIN]. **[syn. nov.]**

    taymansi diffusa DUFRANE, 1952: 24 (*Metarctia*) nom. nov. for *M. t. kamitugensis* DUFRANE, 1945.

    vicina KIRIAKOFF, 1953b: 33-34 (*Hippurarctia*). Holotype [DRC] "Rwankwi" [RMCA]. **[syn. nov.]**

    overlaeti KIRIAKOFF, 1953b: 34-35 (*Hippurarctia vicina* ssp.). Holotype [DRC] „Lualaba: R. Lupweshi" [RMCA]. **[syn. nov.]**

    bergeri KIRIAKOFF, 1953b: 35 (*Hippurarctia*). Holotype [DRC] „Kapanga" [RMCA] (synonymized by Przybyłowicz & Kühne 2008: 151).

50. *judith* KIRIAKOFF, 1959b: 26-28 (*Hippurarctia*). Holotype [DRC] "Uele: Paulis" [RMCA].

51. *taymansi* ROTHSCHILD, 1910: 442 (*Metarctia*). Holotype [DRC] "Kassai District, Congo Free State" [BMNH].

    septentrionalis KIRIAKOFF, 1957c: 97 (*Hippurarctia taymansi* ssp.). Holotype [DRC] "E. Upper Ituri Valley, 30 miles south of Irumu" [BMNH]. **[syn. nov.]**

Lempkeella KIRIAKOFF, 1953b: 76.

Type species: *Apisa dufranei* KIRIAKOFF, 1952 (by original designation).

52. *avellana* KIRIAKOFF, 1957a: 281-282 (*Bergeria*). Holotype [DRC] "Tshuapa: Eala" [RMCA]. **[comb. nov.]**

53. *dufranei* KIRIAKOFF, 1952b: 402-404 (*Apisa*). Holotype [DRC] "Elisabethville" [RMCA].

54. *vanoyei* KIRIAKOFF, 1952b: 404-406 (*Apisa*). Holotype [DRC] "Bolombo" [RMCA].

Mecistorhabdia KIRIAKOFF, 1953b: 29.

Type species: *Metarctia haematoessa* HOLLAND, 1893 (by original designation).

55. *haematoessa* HOLLAND, 1893: 396 (*Metarctia*). Holotype [Gabon] "Valley of the Ogove River" [CMNH].

    burgessi KIRIAKOFF, 1957c: 96-97 (*Mecistorhabdia*). Holotype [Uganda] „Kigezi Dist., Impenetrable Forest, Kanungu" [BMNH] (synonymized by Przybyłowicz & Kühne 2008: 150).

Melisa WALKER, 1854: 264-265.

Type species: *Euchromia diptera* WALKER, 1854 (by monotypy).

56. *croceipes* AURIVILLIUS, 1892: 200 (*Balacra?*). Syntypes [Cameroon] "Camerun" [NHRS]. **[stat. res.]**
    atavistis HAMPSON, 1911: 395 (*Melisa*). Syntypes [Ghana] "Bibianaha" [BMNH]. **[syn. nov.]**
    mariae DUFRANE, 1945: 136 (*Melisa atavistis* f. n.). Holotype [DRC] "Kamituga" [KBIN]. **[syn. nov.]**

57. *diptera* WALKER, 1854: 265 (*Euchromia*). Holotype [DRC or Congo-Brazzaville] „Congo" [BMNH].
    grandis HOLLAND, 1893: 394 (*Melisa*). Holotype [Gabon] „Valley of the Ogove River" [CMNH] (synonymized by Hampson 1898: 152).

58. *hancocki* JORDAN, 1936: 292-293 (*Melisa*). Syntypes [Uganda] „Mabiri Forest, Kololo" [OUM].

Melisoides STRAND, 1912: 192.
Type species: *Melisoides lobata* STRAND, 1912 (by original designation).
    Collartisa KIRIAKOFF, 1953b: 81.
    Type species: *Collartisa collartorum* KIRIAKOFF, 1953b (by original designation).

59. *lobata* STRAND, 1912: 193 (*Melisoides*). Holotype [Equatorial Guinea] "Alen" [ZMHB].
    bitjeana BETHUNE-BAKER, 1927: 321 (*Paramelisa*). Lectotype [Cameroon] "Bitye; Ja River" [BMNH] (synonymized by Przybyłowicz & Dall'Asta 2003: 348).
    collartorum KIRIAKOFF, 1953b: 81-82 (*Collartisa*). Holotype [DRC] "Bokuma" [RMCA] (synonymized by Przybyłowicz & Dall'Asta 2003: 348).

Metamicroptera HULSTAERT, 1923: 408-409.
Type species: *Metamicroptera rotundata* HULSTAERT, 1923 (by original designation).
    Neobalacra KIRIAKOFF, 1952c: 79.
    Type species: *Balacra paradoxa* HERING, 1932. (by monotypy).
    Micrometaptera KIRIAKOFF, 1960: 3, 5, 62 (an incorrect subsequent spelling of *Metamicroptera* HULSTAERT, 1923).

60. *christophi* PRZYBYŁOWICZ, 2005: 140-142 (*Metamicroptera*). Holotype: [Zambia] „Nkana" [TMSA].

61. *rotundata* HULSTAERT, 1923: 409 (*Metamicroptera*). Lectotype [DRC] "Elisabethville" [RMCA].
    paradoxa HERING, 1932: 107 (*Balacra*). Lectotype [DRC] "Lubumbashi" [RMCA] (synonymized by Kiriakoff 1953b: 75).
    paradoxa ROMIEUX, 1934a: 143-144 (*Balacra*). Lectotype [DRC] "Ht Katanga, Tshinkolobwe" [MHNG] (synonymized with *B. paradoxa* HER. by Romieux 1946: 267).

Metarctia WALKER, 1855: 769.
Type species: *Metarctia rufescens* WALKER, 1855 (by monotypy).

Metaretia Pagenstecher, 1909: 420 (an incorrect subsequent spelling of *Metarctia* Walker, 1855).

Subgenus *Collocaliodes* Kiriakoff, 1957b: 145.

Type species: *Collocaliodes dracoena* Kiriakoff, 1953 (by original designation).

62. *aethiops* Kiriakoff, 1973a: 61-62 (*Automolis*). Holotype [Malawi] "Mkuwadzi Forest, Nkata Bay" [BMZ].

63. *collocalia* Kiriakoff, 1957c: 110 (*Metarctia*). Holotype [Zimbabwe] "Vumba" [BMNH].
  kilimaensis Kiriakoff, 1973a: 63 (*Automolis collocalia* ssp.). Holotype [Tanzania] "Kilimanjaro, Marangu" [ZSM].
  montium Kiriakoff, 1957c: 110-111 (*Metarctia collocalia* ssp.). Holotype [DRC] "Rutshuru to Kabali" [BMNH].

64. *debauchei* Kiriakoff, 1953b: 45-46 (*Metarctia*). Holotype [Burundi] „Usumbura" [RMCA]. **[comb. nov.]**

65. *dracoena* Kiriakoff, 1953b: 53-54 (*Metarctia*). Holotype [DRC] „Lusambo" [RMCA].

66. *fuliginosa* Kiriakoff, 1953b: 55 (*Metarctia*). Holotype [DRC] „Lusambo" [RMCA].

67. *jansei* Kiriakoff, 1957c: 109-110 (*Metarctia*). Holotype [RSA] „Natal" [BMNH].

68. *olbrechtsi* Kiriakoff, 1953b: 54-55 (*Metarctia*). Holotype [DRC] „Lubudi s. d." [RMCA].

69. *pavlitzkae* Kiriakoff, 1961b: 107-108 (*Metarctia*). Holotype [Tanzania] „Sakarani" [ZSM].

70. *seydeliana* Kiriakoff, 1953b: 54 (*Metarctia*). Holotype [DRC] „Elisabethville" [RMCA].

71. *tenebrosa* Le Cerf, 1922: 393 (*Metarctia*). Holotype [Kenya or Uganda] "Afrique Orientale Anglaise" [MNHN].
  margaretha Kiriakoff, 1957c: 111 (*Metarctia*). Holotype [Kenya] "Nairobi" [BMNH]. **[syn. nov.]**

Subgenus *Hebena* Walker, 1856: 1722-1723.

Type species: *Hebena venosa* Walker, 1856 (by monotypy).
  Hexaneura Wallengren, 1860: 164 (by original designation).
  Type species: *Hexaneura cinnamomea* Wallengren, 1860.

72. *cinnamomea* Wallengren, 1860: 164 (*Hexaneura*). Holotype [RSA] "Caffraria" [NHRS]. **[stat. res.]**

73. *henrardi* Kiriakoff, 1953b: 42 (*Metarctia lateritia henrardi*). Holotype [DRC] "terr. Kanda-Kanda: Gandajika" [RMCA].

74. *lateritia* Herrich-Schäffer, 1850-1858: 72, 81, fig. 274 (*Automolis*). Syntype(s) [RSA] „Cap." [MLU - not located (pers. comm. from J. Haendel)].
  venosa Walker, 1856: 1723 (*Hebena*). Holotype [RSA] "Port Natal" [OUM] (synonymized by Hampson 1898: 145).

75. *rubra* WALKER, 1856: 1720-1721 (*Anace*). Lectotype **here designated** [RSA] "Cape G. Hope" [BMNH]. **[comb. nov.]**

    kelleni SNELLEN, 1886a: 1-2 (*Automolis*). Syntypes [Angola] "Humpata" [RNHL]. **[syn. nov.]**

    titan TALBOT, 1929: 73 (*Metarctia*). Holotype [Angola] "Upper Cubango-Cunene Watershed" [BMNH]. **[syn. nov.]**

76. *subincarnata* KIRIAKOFF, 1954a: 1-2 (*Metarctia*). Holotype [DRC] „Lupweji" [KBIN].

Subgenus *Metarctia* WALKER, 1855: 769.

Type species: *Metarctia rufescens* WALKER, 1855 (by monotypy).

    Notharctia KIRIAKOFF, 1953b: 42.

    Type species: *Notharctia flavivena* HAMPSON, 1901 (by original designation).

    Oenarctia KIRIAKOFF, 1953b: 51. **[syn. nov.]**

    Type species: *Oenarctia erlangeri* ROTHSCHILD, 1910 (by original designation).

    Pterophaea KIRIAKOFF, 1953b: 53.

    Type species: *Pterophaea brunneipennis* HERING, 1932 (by original designation).

77. *alticola* AURIVILLIUS, 1925a: 11 (*Metarctia rufescens* (ab?.) *alticola* n. var.). Syntypes [Rwanda] "Rwanda District und Birunga" [NHRS]. **[stat. res.]**

    rhodites KIRIAKOFF, 1957c: 106-107 (*Metarctia*). Holotype [Rwanda] „Rugege Forest, Rwanda dist., Lake Kivu" [BMNH]. **[syn. nov.]**

78. *atrivenata* KIRIAKOFF, 1956b: 39 (*Metarctia*). Holotype [Tanzania] „12 ml E of Mbeya" [RMCA].

79. *benitensis* HOLLAND, 1893: 395 (*Metarctia*). Syntypes [Equatorial Guinea], "Benita" [CMNH].

    paniscus KIRIAKOFF, 1957a: 279 (*Metarctia*). Holotype [DRC] "Kibali-Ituri: Nioka" [RMCA]. **[syn. nov.]**

    subnigra KIRIAKOFF, 1958: 43-44 (*Metarctia*). Holotype [Uganda] „Mahoma River" [BMNH]. **[syn. nov.]**

80. *brunneipennis* HERING, 1932: 108 (*Metarctia*). Holotype [DRC] "Elisabethville" [RMCA].

81. *burra* SCHAUS in SCHAUS & CLEMENTS, 1893: 23 (*Anace*). Holotype [Sierra Leone] [AMNH].

    hector KIRIAKOFF, 1959b: 30 (*Metarctia*). Holotype [DRC] "Uele: Paulis" [RMCA]. **[syn. nov.]**

    chryseis KIRIAKOFF, 1973a: 57-58 (*Automolis*). Holotype [Nigeria] "Ogoja Co." [ZSM]. **[syn. nov.]**

    sudanica KIRIAKOFF, 1973b: 90-91 (*Automolis*). Holotype [Sudan] "Prov. Kordofan, Kadugli" [ZSM]. **[syn. nov.]**

82. *burungae* DEBAUCHE, 1942: 11 (*Metarctia*). Holotype [DRC] „Burunga (Mokoto)" [RMCA].

    umbretta KIRIAKOFF, 1963a: 95 (*Metarctia*). Holotype [DRC] "Kyandolire, camp des gardes" [RMCA]. **[syn. nov.]**

83. *carmel* Kiriakoff, 1957c: 105 (*Metarctia*). Holotype [Ethiopia] "Kambatta" [BMNH].

84. *diversa* Bethune-Baker, 1911: 532-533 (*Metarctia*). Holotype [Angola] "N'Dalla Tando" [BMNH].   **[stat. rev.] [comb. nov.]**

    pallidicosta Hulstaert, 1923: 406-407 (*Metarctia*). Syntypes [DRC] "Elisabethville, Nieuwdorp" [RMCA]. **[syn. nov.]**

85. *fario* Kiriakoff, 1957c: 103-104 (*Metarctia*). Holotype [DRC] "Katenge, River Katay" [BMNH].

86. *flaviciliata* Hampson, 1907: 225 (*Metarctia*). Lectotype **here designated** [DRC] "Beni Semliki" [BMNH].

87. *flavicincta* Aurivillius, 1900: 1057 (*Metarctia*). Holotype [DRC] "Congo" [NHRS].

    contrasta Bethune-Baker, 1911: 533 (*Metarctia*). Holotype [Angola] "N'Dalla Tando" [BMNH] (synonymized by Hampson 1914a: 70).

    jacksoni Kiriakoff, 1956b: 40 (*Metarctia*). Holotype [Uganda] „Madi Opei, N. Acholi" [RMCA] (synonymized by Przybyłowicz & Kühne 2008: 152).

    dracuncula Kiriakoff, 1957a: 277-278 (*Metarctia*). Holotype [DRC] „Congo Belge: Felitebwe" [RMCA] (synonymized by Przybyłowicz & Kühne 2008: 152).

    subpumila Kiriakoff, 1957a: 279 (*Metarctia*). Holotype [Angola] „Kabinda" [RMCA] (synonymized by Przybyłowicz & Kühne 2008: 152).

    moira Kiriakoff, 1957c: 109 (*Metarctia*). Holotype [Sudan] „Tembura" [BMNH] (synonymized by Przybyłowicz & Kühne 2008: 152).

88. *flavivena* Hampson, 1901: 169 (*Metarctia*). Syntypes [Kenya] "Machakos" [BMNH].

    flavivena Hampson, 1902: 40 (*Metarctia*). **Second print of 1901 description.**

    zegina Strand, 1920: 224 (*Metarctia flavivena* n. var.). Syntypes "Abyssinien" [BMNH] - a subspecific name for *M. flavivena* Subsp. 1. in Hampson 1914a: 70. **[syn. nov.]**

    panyamana Strand, 1920: 224 (*Metarctia flavivena* n. var.). Syntypes "N. Nigeria" [BMNH] - a subspecific name for *M. flavivena* Subsp. 2. in Hampson 1914a: 70. **[syn. nov.]**

    rothschildi Le Cerf, 1922: 392 (*Metarctia*). Holotype [Ethiopia] "Parages du mont Assabot" [MNHN] (synonymized with *M. zegina* by Kiriakoff 1957b: 133).

89. *flora* Kiriakoff, 1957c: 105-106 (*Metarctia*). Holotype [Rwanda] "Rugege Forest, Rwanda dist., Lake Kivu" [BMNH].

    katriona Kiriakoff, 1957c: 106 (*Metarctia*). Holotype [Rwanda] "Rugege Forest, Rwanda dist., Lake Kivu" [BMNH]. **[syn. nov.]**

90. *fontainei* Kiriakoff, 1953b: 26 (*Metarctia flavicincta fontainei*). Holotype [DRC] "Katako-Kombe" [RMCA]. **[stat. res.]**

91. *forsteri* Kiriakoff, 1955b: 259-260 (*Metarctia*). Holotype „Kamerun" [ZSM].

92. *fulvia* Hampson, 1901: 170 (*Metarctia*). Lectotype **here designated** [Kenya] "Athi ya Mawe" [BMNH].

    neaera Fawcett, 1915: 92 (*Metarctia*). Lectotype **here designated** [Kenya] "Kedai" [BMNH]. **[syn. nov.]`**

93. *fusca* HAMPSON, 1901: 169-170 (*Metarctia*). Holotype [Kenya] "Kikuju, Romoro" [BMNH].

94. *galla* ROUGEOT, 1977: 82 (*Metarctia*). Holotype [Ethiopia] "reserve de Balé" [MNHN].

95. *haematricha* HAMPSON, 1905: 426 (*Metarctia*). Holotype [Ethiopia] "Kutai Mecha" [BMNH].
latipennis KIRIAKOFF, 1957c: 102-103 (*Metarctia*). Holotype [Ethiopia] "Dangila" [BMNH]. **[syn. nov.]**

96. *hebenoides* KIRIAKOFF, 1973a: 56 (*Automolis*). Holotype [Malawi] "Chikangawa" [BMZ].

97. *hulstaertiana* KIRIAKOFF, 1953b: 46 (*Metarctia*). Holotype [DRC] "Bokote" [RMCA].

98. *inconspicua* HOLLAND, 1892: 93 (*Metarctia*). Holotype [Tanzania] "Zanzibar" [USNM].
inconspicua HOLLAND, 1896: 250 (*Metarctia*). **Second print of 1892 description.**

99. *johanna* KIRIAKOFF, 1979: 238-239 (*Automolis*). Holotype [Nigeria] "Ogoja" [ZSM].

100. *kumasina* STRAND, 1920: 224 (*Metarctia pallida* var. *kumasina*). Lectotype [Ethiopia] "Abyssinia, Zegi Tsana" [BMNH]. **[stat. res.].**
jubdoensis KIRIAKOFF, 1955b: 261-262 (*Metarctia*). Holotype [Ethiopia] "Jubdo, Wallega" [ZSM]. **[syn. nov.]**

101. *lindemannae* KIRIAKOFF, 1961b: 104-105 (*Metarctia*). Holotype [Tanzania] „Sakarani" [ZSM].

102. *longipalpis* HULSTAERT, 1923: 406 (*Metarctia*). Holotype [DRC] "Elisabethville" [RMCA].

103. *lugubris* GAEDE, 1926: 118 (*Metarctia*). Holotype [Tanzania] „Ukingaberge" [ZMHB].
usta DEBAUCHE, 1942: 10 (*Metarctia*). Holotype [DRC] „Parc National Albert, lac Magera" [RMCA]. **[syn. nov.]**
fletcheri KIRIAKOFF, 1958: 44-45 (*Metarctia*). Holotype [Uganda] „Nyamaleju" [BMNH]. **[syn. nov.]**

104. *maria* KIRIAKOFF, 1957c: 107-109 (*Metarctia*). Holotype [Guinea] „Boukouni near Macenta" [BMNH].

105. *metaleuca* HAMPSON, 1914a: 69 (*Metarctia*). Holotype [Liberia] "Nanna Kru " [BMNH].
pamela KIRIAKOFF, 1957c: 104 (*Metarctia*). Holotype [Cameroon] „Bitje, Ja river" [BMNH]. **[syn. nov.]**

106. *morag* KIRIAKOFF, 1957c: 102 (*Metarctia*). Holotype [DRC] "Upper Oso River, NW Kivu" [BMNH].
hecqi KIRIAKOFF, 1959b: 32 (*Metarctia*). Holotype [DRC] "Kibali-Ituri: Nioka" [RMCA]. **[syn. nov.]**

107. *negusi* KIRIAKOFF, 1957c: 100 (*Metarctia crassa negusi*). Holotype [Ethiopia] „Abyssinia" [BMNH]. **[stat. res.]**

108. *nigritarsis* BERIO, 1943: 176 (*Metarctia pallida* (?) f. *nigritarsis* nov.). Holotype [Eritrea] "Dorfu" [MCSNG]. **[stat. res.]**

109. *noctis* Druce, 1910: 394 (*Metarctia*). Holotype [Ethiopia] „Dire Daoua" [BMNH].

110. *pallens* Bethune-Baker, 1911: 532 (*Metarctia*). Holotype [Angola] „N'Dalla Tando" [BMNH].

111. *paremphares* Holland, 1893: 395 (*Metarctia*). Syntypes [Gabon] "Valley of the Ogove River" [CMNH].
  orientalis Kiriakoff, 1956b: 39-40 (*Metarctia benitensis* ssp.). Holotype [Kenya] "Mt. Elgon" [RMCA] (synonymized by Przybyłowicz & Kühne 2008: 153).
  capricornis Kiriakoff, 1957c: 103 (*Metarctia*). Holotype [Angola] "Fazende Congulu, Amboim Dist." [BMNH] (synonymized by Przybyłowicz & Kühne 2008: 153).

112. *paulis* Kiriakoff, 1961a: 10 (*Metarctia*). Holotype [DRC] "Uele: Paulis" [RMCA].

113. *phaeoptera* Hampson, 1909: 344 (*Metarctia*). Holotype [DRC] "Upper Congo" [BMNH].
  xanthippa Kiriakoff, 1956b: 41-42 (*Metarctia*). Holotype [Uganda] "Madi Opei, Acholi" [RMCA]. **[syn. nov.]**
  helga Kiriakoff, 1957c: 107 (*Metarctia*). Holotype [Tanzania] „Tanganyika Territory, Kigoma dist., Buswazi, Kasulu" [BMNH]. **[syn. nov.]**
  tricolor Rougeot, 1977: 82 (*Metarctia*). Holotype [Ethiopia] "Riv. Baro" [MNHN]. **[syn. nov.]**

114. *priscilla* Kiriakoff, 1957c: 104 (*Metarctia*). Holotype [Ghana] "Bibianaha, 70 miles N.W. of Dimkwa" [BMNH].

115. *pulverea* Hampson, 1907: 225 (*Metarctia*). Holotype [Uganda] "Ruwenzori, 6000'" [BMNH].
  bipuncta Joicey & Talbot, 1924: 549 (*Metarctia*). Syntypes [Rwanda] "Rugege Forest" [BMNH] (synonymized by Kiriakoff 1956b: 41).

116. *pumila* Hampson, 1909: 344-345 (*Metarctia*). Lectotype **here designated** [Sudan] "White Nile, Gondokoro" [BMNH].

117. *robusta* Kiriakoff, 1973a: 55-56 (*Automolis*). Holotype [Zambia] „Mbala, Sambien" [BMZ - not located (pers. comm. from V. Muyambo)].

118. *rufescens* Walker, 1855a: 769-770 (*Metarctia*). Syntypes [RSA] "Caffraria; Port Natal" [BMNH].
  maculifera Wallengren, 1860: 164 (*Hexaneura*). Holotype [RSA] "Caffraria" [NHRS] (synonymized by Hampson 1898: 148).

119. *saalfeldi* Kiriakoff, 1960b: 42 (*Metarctia burra* ssp.). Holotype [Ethiopia] "Villagio" [ZSM]. **[stat. res.]**

120. *salmonea* Kiriakoff, 1957c: 103 (*Metarctia*). Holotype [Angola] "Dondo" [BMNH].

121. *sarcosoma* Hampson, 1901: 170 (*Metarctia*). Holotype [Kenya] "Machakos" [BMNH].

122. *sheljuzhkoi* Kiriakoff, 1961b: 105-106 (*Metarctia*). Holotype [Ivory Coast] „Abidjan" [ZSM].

123. subpallens KIRIAKOFF, 1956b: 38 (*Metarctia*). Holotype [Kenya] „Makueni" [RMCA].

124. tenera KIRIAKOFF, 1973a: 59-60 (*Automolis*). Holotype [Zimbabwe] „Mchabezi Valley, Matopos, Bulawayo" [BMZ].

125. transvaalica KIRIAKOFF, 1973a: 60-61 (*Automolis*). Holotype [RSA] „Pretoria" [BMZ].

126. tricolorana WICHGRAF, 1922: 171 (*Metorctia* [sic]). Holotype [Uganda] „Gulu" [BMNH]. **[stat. rev.]**

127. unicolor OBERTHÜR, 1880: 186 (*Automolis*). Holotype [Ethiopia] "Presa in viaggio venando da Fin-Finni paese dei Galla" [MCSNG]. **[stat. rev.]** **[comb. nov.]**

> *erlangeri* ROTHSCHILD, 1910: 442 (*Metarctia*). Holotype [Ethiopia] "Djedda to Sibbe, Arussi Galla Country, Abyssinia" [BMNH]. **[syn. nov.]**
>
> *major* LE CERF, 1922: 393-394 (*Metarctia*). Holotype [Ethiopia] "Addis-Abeba" [MNHN]. **[syn. nov.]**
>
> *aegrota* BERIO, 1939: 48 (*Metarctia lateritia* n. f. *aegrota*). Holotype [Eritrea] "Elaberet" [MCSNG] (synonymized with *M. major* by Kiriakoff 1960b: 37).
>
> *abyssinibia* KIRIAKOFF, 1957c: 100-102 (*Metarctia*). Holotype [Ethiopia] "Abyssinia: Abbai Affat" [BMNH] (synonymized with *M. major* by Kiriakoff 1960b: 37).

128. uniformis BETHUNE-BAKER, 1911: 534 (*Metarctia*). Holotype [Angola] "Malange" [BMNH].

129. upembae KIRIAKOFF, 1954b: 28 (*Metarctia*). Holotype [DRC] „Kankunda, rive dr[oit]. Lupiala" [RMCA]. **[comb. nov.]**

130. venustissima KIRIAKOFF, 1961a: 10-11 (*Metarctia*). Holotype [DRC] „Katanga, Kolwezi" [RMCA].

131. virgata JOICEY & TALBOT, 1921: 158 (*Metarctia*). Syntypes [DRC] "Mikeno Volcano" [BMNH].
> *wittei* DEBAUCHE, 1942: 11-12 (*Metarctia*). Holotype [DRC] " Parc National Albert, riviere Bishakishaki (Kamatembe)" [RMCA] (synonymized by Kiriakoff 1956b: 41).

Subgenus *Metarhodia* KIRIAKOFF, 1953b: 36.
Type species: *Metarctia rubripuncta* HAMPSON, 1898 (by original designation).

132. confederationis KIRIAKOFF, 1961b: 101 (*Metarctia*). Holotype [RSA] "Natal, Karkloof" [ZSM].

133. epimela KIRIAKOFF, 1979: 237-238 (*Metarhodia*). Holotype [Tanzania] „Mt. Meru, Momella" [ZSM].

134. heinrichi KIRIAKOFF, 1961b: 100 (*Metarctia*). Holotype [Angola] „Canzele" [ZSM].

135. heringi KIRIAKOFF, 1957a: 275 (*Metarctia*). Holotype [DRC] "Elisabethville" [RMCA].

136. hypomela KIRIAKOFF, 1956a: 24 (*Metarctia*). Holotype [Kenya] „Kakamega" [RMCA].
> *kenyae* KIRIAKOFF, 1957c: 97 (*Metarctia*). Holotype [Kenya] „Nekuru" [BMNH] (synonymized by Przybyłowicz & Kühne 2008: 151).

137. *insignis* KIRIAKOFF, 1959b: 28 (*Metarctia*). Holotype [Rwanda] „Kisenyi" [RMCA].

138. *jordani* KIRIAKOFF, 1957c: 98 (*Metarctia*). Holotype [Angola] „Mt. Moco, Luimbale" [BMNH].

139. *nigricornis* DEBAUCHE, 1942: 9-10 (*Metarctia haematosphages nigricornis*). Holotype [DRC] "Nyarusambo (Mikeno)" [RMCA]. **[stat. res.]**

140. *rubribasa* BETHUNE-BAKER, 1911: 533 (*Metarctia*). Holotype [Angola] "N'Dalla Tando" [BMNH]. **[comb. nov.]**

> *rubricosta* TALBOT, 1929 (*Metarctia rubripuncta* f. nov. *rubricosta*). Holotype [Angola] "Upper Cubango-Cunene Watershed" [BMNH]. **[comb. nov.]** **[syn. nov.]**

> *deriemaeckeri* KIRIAKOFF, 1953b: 37-38 (*Metarctia*). Holotype [DRC] „Elisabethville" [RMCA]. **[syn. nov.]**

141. *rubripuncta* HAMPSON, 1898: 147 (*Metarctia*). Holotype "Gabon" [BMNH].

> *chapini* HOLLAND, 1920: 264 (*Metarctia*). Holotype [DRC] "Medje" [AMNH] (synonymized by Przybyłowicz & Kühne 2008: 151).

> *denisae* DUFRANE, 1945: 131-132 (*Metarctia*). Holotype [DRC] „Kamituga" [KBIN] (synonymized by Kiriakoff 1953b: 37).

> *impura* KIRIAKOFF, 1959b: 28 (*Metarctia*). Holotype [DRC] „Uele: Paulis" [RMCA] (synonymized by Przybyłowicz & Kühne 2008: 151).

Subgenus *Pinheyata* NYE in WATSON, FLETCHER & NYE, 1980: 154.

Type species: *Metarctia quinta* KIRIAKOFF, 1973 (by original designation for *Pinheya*).

> *Pinheya* KIRIAKOFF, 1973a: 63.

> Type species: *Metarctia quinta* KIRIAKOFF, 1973 (by original designation).

142. *crocina* KIRIAKOFF, 1973a: 65-66 (*Automolis*). Holotype [Zimbabwe] „Tobacco Res[earch] St[atio]n, Salisbury" [BMZ].

143. *quinta* KIRIAKOFF, 1973a: 64-65 (*Automolis*). Holotype [Zimbabwe] "Vumba Mts., Umtali" [BMZ].

Subgenus *Thyretarctia* STRAND, 1912: 189.

Type species: *Metarctia haematica* HOLLAND, 1893 (by monotypy).

144. *brunneoaurantiaca* KIRIAKOFF, 1973a: 53-54 (*Automolis*). Holotype [Kenya] "Isiola" [BMZ].

145. *didyma* KIRIAKOFF, 1957c: 98 (*Metarctia*). Holotype [Ghana] „Kumasi" [BMNH].

146. *haematica* HOLLAND, 1893: 396 (*Metarctia*). Syntypes [Gabon] "Kangwe, Valley of the Ogove River" [CMNH].

> *haematosphages* HOLLAND, 1893: 396 (*Metarctia*). Holotype [Gabon] "Valley of the Ogove River" [CMNH] (synonymized by Przybyłowicz & Kühne 2008: 152).

147. *infausta* KIRIAKOFF, 1957a: 276 (*Metarctia*). Holotype [DRC] „Kibali-Ituri: Nioka" [RMCA].

148. *morosa* KIRIAKOFF, 1957a: 276 (*Metarctia*). Holotype [DRC] „Haut-Katanga: Elisabethville" [RMCA].

149. *schoutedeni* KIRIAKOFF, 1953b: 39-40 (*Metarctia*). Holotype [DRC] „Kivu: Burunga" [RMCA].

pinheyi KIRIAKOFF, 1956a: 24 (*Metarctia*). Holotype [Kenya] „Kitale" [RMCA] (synonymized by Przybyłowicz & Kühne 2008: 152).

Microbergeria KIRIAKOFF, 1972: 102-103.

Type species: *Microbergeria luctuosa* KIRIAKOFF, 1972: 104 (by original designation).

150. *luctuosa* KIRIAKOFF, 1972: 104 (*Microbergeria*). Holotype [Cameroon] "Lomie" [BMZ].

Neophemula KIRIAKOFF, 1957a: 282.

Type species: *Pseudapiconoma vitrina* OBERTHÜR, 1909 (by original designation).

151. *vitrina* OBERTHÜR, 1909: 95 (*Pseudapiconoma*). Holotype [Cameroon] "Kamerun" [BMNH].

congoensis KIRIAKOFF, 1957a: 283 (*Neophemula vitrina* ssp.). Holotype [DRC] "Congo Belge: Fomhoko" [RMCA].

angolensis KIRIAKOFF, 1957c: 113 (*Neophemula vitrina* ssp.). Holotype [Angola] "Quicolungo, 120 km N of Lucala" [BMNH].

Owambarctia KIRIAKOFF, 1957a: 274.

Type species: *Owambarctia owamboensis* KIRIAKOFF, 1957 (by original designation).

152. *owamboensis* KIRIAKOFF, 1957a: 274-275 (*Owambarctia*). Holotype [Namibia] "Tsumeb, région de l'Owambo" [RMCA].

153. *unipuncta* KIRIAKOFF, 1973b: 88-89 (*Owambarctia*). Holotype [Tanzania] "Uruguru-Berge" [ZSM].

Paramelisa AURIVILLIUS, 1905b: 13.

Type species: *Paramelisa lophura* AURIVILLIUS, 1905 (by monotypy).

154. *dollmani* HAMPSON, 1920: 253 (*Paramelisa*). Lectotype **here designated** [Zambia] "Solwezi" [BMNH].

155. *leroyi* KIRIAKOFF, 1953b: 80-81 (*Paramelisa*). Holotype [DRC] „Rwankwi" [RMCA].

156. *lophura* AURIVILLIUS, 1905b: 13 (*Paramelisa*). Holotype [DRC] "Mukimbungu" [NHRS].

157. *lophuroides* OBERTHÜR, 1911: 468 (*Paramelisa*). Holotype [Cameroon] „Johann Albrechts Höhe Station" [BMNH].

Pseudmelisa HAMPSON, 1910: 391.

Type species: *Pseudmelisa chalybsa* HAMPSON, 1910 (by original designation).
    Pseudomelisa ZERNY, 1912a: 47 (an unjustified emendation).

158. *chalybsa* HAMPSON, 1910: 391 (*Pseudmelisa*). Holotype [DRC] "Kambove distr" [BMNH].
    chalybea ZERNY, 1912a: 47 (an unjustified emendation).

159. *rubrosignata* KIRIAKOFF, 1957c: 114 (*Pseudmelisa*). Holotype [Malawi] "Nyasaland, Mlanje, Luchenya river" [BMNH].

Pseudothyretes DUFRANE, 1945: 127.

    Type species: *Pseudothyretes mariae* DUFRANE, 1945 (by original designation).
    Diakonoffia KIRIAKOFF, 1953b: 22 (synonymized by Przybyłowicz & Kühne 2008: 154).
    Type species: *Apisa kivensis* DUFRANE, 1945 (by original designation).

160. *carnea* HAMPSON, 1898: 136 (*Meganaclia*). Holotype "Angola" [BMNH]. **[stat. rev.]**

161. *erubescens* HAMPSON, 1901: 169 (*Tritonaclia*). Holotype [Uganda] "Uganda Ry., mile 478" [BMNH].

162. *kamitugensis* DUFRANE, 1945: 128-129 (*Apisa*). Holotype [DRC] "Kamituga" [KBIN].

163. *mariae* DUFRANE, 1945: 127 (*Pseudothyretes*). Holotype [DRC] „Kamituga" [KBIN]. **[stat. rev.]**

164. *nigrita* KIRIAKOFF, 1961a: 9 (*Diakonoffia*). Holotype [DRC] „Uele: Paulis" [RMCA].

165. *perpusilla* WALKER, 1856: 1720 (*Anace*). Holotype [no locality] [BMNH].

166. *rubicundula* STRAND, 1912: 187 (*Metarctia*). Holotype [Equatorial Guinea] "Makomo (Ntumgebiet)" [ZMHB].
    kivensis DUFRANE, 1945: 128 (*Apisa*). Holotype [DRC] „Kamituga" [KBIN] (synonymized by Kiriakoff 1960b: 48).

Rhabdomarctia KIRIAKOFF, **1953b: 30.**

Type species: *Metarctia rubrilineata* BETHUNE-BAKER, 1911 (by original designation).

167. *rubrilineata* BETHUNE-BAKER, 1911: 533 (*Metarctia*). Holotype [Angola] "N'Dalla Tando" [BMNH].

    ochreogaster JOICEY & TALBOT, 1921: 158-159 (*Metarctia*). Lectotype [DRC] "Ituri Forest, N.W. Beni" [BMNH] (synonymized by Kiriakoff 1960b: 28).

    waelbroecki DEBAUCHE, 1938: 1 (*Metarctia*). Holotype [DRC] "Kinchassa" [KBIN] (synonymized by Kiriakoff 1960b: 28).

    kamitugensis DUFRANE, 1945: 130-131 (*Metarctia*). Holotype [DRC] "Kamituga" [KBIN] (synonymized by Przybyłowicz & Kühne 2008: 150).

    similis KIRIAKOFF, 1953b: 31-32 (*Rhabdomarctia*). Holotype [DRC] "Eala" [RMCA] (synonymized by Przybyłowicz & Kühne 2008: 150).

    bipartita KIRIAKOFF, 1973a: 51 (*Rhabdomarctia*). Holotype [Tanzania] „Bukoba" [ZSM] (synonymized by Przybyłowicz & Kühne 2008: 150).

    costalis KIRIAKOFF, 1973b: 88 (*Rhabdomarctia*). Holotype [Uganda] "Kalinzu Forest" [ZSM] (synonymized by Przybyłowicz & Kühne 2008: 150).

Rhipidarctia KIRIAKOFF, **1953b: 28.**

Type species: *Elsa rubrosuffusa* KIRIAKOFF, 1953b (by original designation).

Subgenus *Elsitia* PRZYBYŁOWICZ [**nom. nov.**]

Type species: *Metarctia pareclecta* HOLLAND, 1893

168. *cinctella* KIRIAKOFF, 1953b: 56 (*Metarctia*). Holotype [DRC] "Lusambo" [RMCA].

    strenua KIRIAKOFF, 1957a: 273 (*Rhipidarctia*). Holotype [DRC] "Sankuru: Katako-Kombe" [RMCA]. **[syn. nov.]**

169. *forsteri* KIRIAKOFF, 1953b: 27-28 (*Elsa*). Holotype [DRC] „Rwankwi" [RMCA].

    unicolor KIRIAKOFF, 1957a: 272-273 (*Rhipidarctia*). Holotype [DRC] "Tshuapa: Eala" [RMCA] (synonymized by Przybyłowicz & Kühne 2008: 150).

    ruandae KIRIAKOFF, 1961a: 9-10 (*Rhipidarctia forsteri* ssp.). Holotype [Rwanda] „Kisenyi" [RMCA]. **[syn. nov.]**

    punctulata KIRIAKOFF, 1963a: 94 (*Rhipidarctia*). Holotype [DRC] „Kyandolire, camp des gardes" [RMCA] (synonymized by Przybyłowicz & Kühne 2008: 150).

170. *lutea* HOLLAND, 1893: 396 (*Metarctia*). Holotype [Gabon] "Valley of the Ogove River" [USNM]

    ghesquierei KIRIAKOFF, 1953b: 27 (*Elsa [invaria] lutea ghesquierei*). Holotype [DRC] "Eala" [RMCA]. **[syn. nov.]**

171. *pareclecta* HOLLAND, 1893: 395-396 (*Metarctia*). Holotype [Gabon] "Valley of the Ogove River" [CMNH].

    rosacea BETHUNE-BAKER, 1911: 532 (*Metarctia*). Holotype [Angola] "N'Dalla Tando" [BMNH] (synonymized by Przybyłowicz & Kühne 2008: 149).

172. *saturata* KIRIAKOFF, 1957a: 273-274 (*Rhipidarctia*). Holotype [DRC] „Sankuru: Katako-Kombe" [RMCA].

173. *subminiata* KIRIAKOFF, 1959b: 25-26 (*Rhipidarctia*). Holotype [DRC] „Uele: Paulis" [RMCA].

Subgenus *Hemirhipidia* KIRIAKOFF, 1955b: 256 .

Type species: *Rhipidarctia danieli* KIRIAKOFF, 1955b (by original designation).

174. *postrosea* ROTHSCHILD, 1913: 187 (*Metarctia*). Holotype [Nigeria] "Near Lagos, 1 mile from Oni [Camp]" [OUM]. **[comb. nov.]**

> *danieli* KIRIAKOFF, 1955b: 256-258 (*Rhipidarctia*). Holotype [Cameroon] "Debundscha" [ZSM]. **[syn. nov.]**

Subgenus *Rhipidarctia* KIRIAKOFF, 1953b: 28.

Type species: *Elsa rubrosuffusa* KIRIAKOFF, 1953b (by original designation).

> *Elsita* KIRIAKOFF, 1954c: 29 (nom. nov. pro *Elsa* KIRIAKOFF, 1953). **[syn. nov.]**
>
> *Elsa* KIRIAKOFF, 1953b: 23 [nec HONRATH, 1892].
>
> Type species: *Anace invaria* WALKER, 1856 (by original designation).
>
> *Takwa* KIRIAKOFF, 1957c: 96. **[syn. nov.]**
>
> Type species: *Takwa xenops* KIRIAKOFF, 1957 (by original designation).

175. *aurora* KIRIAKOFF, 1957a: 269-270 (*Rhipidarctia*). Holotype [DRC] „Sankuru: Katako-Kombe" [RMCA].

176. *conradti* OBERTHÜR, 1911: 468-469 (*Metarctia erubescens* var. *conradti*). Syntype(s) [Cameroon] „Johann-Albrechts Höhe" [BMNH].

> *rhodospila* KIRIAKOFF, 1957a: 270-271 (*Rhipidarctia*). Holotype [DRC] „Sankuru: Ifuta" [RMCA]. **[syn. nov.]**

177. *crameri* KIRIAKOFF, 1961b: 98-99 (*Rhipidarctia*). Holotype [Uganda] „Masindi" [ZSM].

178. *flaviceps* HAMPSON, 1898: 147 (*Metarctia*). Holotype "Cameroons" [BMNH].

> *pallidipes* AURIVILLIUS, 1925b: 1302 (*Metarctia flaviceps* var.). Holotype [Equatorial Guinea: Bioko] "Fernando Poo, Bococo" [ZIMH – destroyed during the Second World War (pers. comm. from H. Riefenstahl)]. **[syn. nov.]**
>
> *rubrosuffusa* KIRIAKOFF, 1953b: 28-29 (*Elsa*). Holotype [DRC] "Eala" [RMCA]. **[syn. nov.]**
>
> *cornelia* KIRIAKOFF, 1957a: 271-272 (*Rhipidarctia*). Holotype [DRC] „Equateur: Flandria" [RMCA]. **[syn. nov.]**

179. *invaria* WALKER, 1856: 1720 (*Anace*). Holotype "West Africa" [BMNH].

> *erubescens* WALKER, 1864: 315 (*Metarctia*). Holotype "Sierra Leone" [BMNH] (synonymized by Hampson 1898: 146).
>
> *melinos* MABILLE, 1890: 37 (*Thyretes*). Holotype [Ivory Coast] "territoire d'Assinie" [BMNH] (synonymized by Hampson 1898: 146).
>
> *aurantiifusca* ROTHSCHILD, 1913: 187 (*Metarctia*). Holotype [Nigeria] "Lagos" [OUM] (synonymized by Kiriakoff 1960b: 25).

180. *miniata* KIRIAKOFF, 1957c: 95-96 (*Rhipidarctia*). Holotype [Cameroon] „Bitje, Ja river" [BMNH].

181. *rhodosoma* KIRIAKOFF, 1957a: 271 (*Rhipidarctia*). Holotype [DRC] „Sankuru: Katako-Kombe" [RMCA].

182. *xenops* KIRIAKOFF, 1957c: 96 (*Takwa*). Holotype [Ghana] "Takwa" [BMNH].

Rhipidarctia INCERTAE SEDIS

183. *rubrovitta* AURIVILLIUS, 1904: 31 (*Metarctia*). Syntypes "Cameroon" [NHRS].

184. *silacea* PLÖTZ, 1880: 86 (*Plegapteryx*). Holotype [Congo] "Abo" [UGD].

185. *syntomia* PLÖTZ, 1880: 85 (*Plegapteryx*). Holotype [Ivory Coast] „Eningo" [UGD].

Thyretes BOISDUVAL, 1847: 596.

Type species: *Thyretes montana* BOISDUVAL, 1847 (by subsequent designation by KIRBY 1892: 103).
    Eressades BETHUNE-BAKER, 1911: 531.
    Type species: *Eressades flavipunctata* BETHUNE-BAKER, 1911 (by original designation).

186. *buettikeri* WILTSHIRE, 1983: 296-297 (*Thyretes*). Holotype [Saudi Arabia] „Feyfa" [NHMB].

187. *caffra* WALLENGREN, 1863: 138 (*Thyretes*). Holotype [RSA, Namibia] „in Caffraria orientali et in territorio fluminis Kuisip" [NHRS].

188. *cooremani* KIRIAKOFF, 1953b: 18 (*Thyretes*). Holotype [DRC] "Léopoldville" [RMCA].

189. *hippotes* CRAMER, [1775-80]: 166, 175, pl. 286 A (*Sphinx*). Syntype(s) [Angola] "Kuft van Bengalen" [BMNH - not located (pers. comm. from M. Honey)].

190. *montana* BOISDUVAL, 1847: 597 (*Thyretes*). Syntype(s) [RSA] "Pays des Amazoulous et de Massilicatzi" [BMNH - not located (pers. comm. from M. Honey)].

191. *monteiroi* BUTLER, 1876: 359 (*Thyretes*). Holotype [Angola] "Ambriz" [BMNH].
    flavipunctata BETHUNE-BAKER, 1911: 531 (*Eressades*). Holotype [Angola] "N'Dalla Tando" [BMNH] (synonymized by Kiriakoff 1957b: 123).
    angolensis GAEDE, 1926: 117 (*Thyretes*). Holotype [Angola] "Malange" [ZMHB] (synonymized by Kiriakoff 1957b: 123).

192. *negus* OBERTHÜR, 1878: 31-32 (*Thyretes*). Holotype [Ethiopia] "Abyssinie" [BMNH].
    phasma BUTLER, 1897: 846 (*Thyretes*). Holotype [Malawi] "Deep Bay" [BMNH] (synonymized by Hampson 1898: 139).
    misa STRAND, 1911: 146-147 (*Thyretes*). Lectotype **here designated** [Togo] "Misahöhe" [ZMHB] (synonymized by Hampson 1914a: 61).

193. *signivenis* HERING, 1937: 229 (*Thyretes*). Holotype [DRC] "Elisabethville" [RMCA].

Thyretini INCERTAE SEDIS

Pseudmelisa HAMPSON, 1910: 391.

194. *demiavis* KAYE, 1919: 88 (*Pseudomelisa* [sic]). Holotype [Cameroon] „Bitje, Ja River" [BMNH].

UNAVAILABLE NAMES

This chapter lists all unavailable names in Thyretini (Arctiidae: Syntominae). These names have no standing in nomenclature; they cannot enter into synonymy and so they are listed separately from the main catalogue.

lateritia ab. **abyssinibia** STRAND, 1920: 224 (*Metarctia*). Syntypes "Abyssinia" [BMNH] - an infrasubspecific name for *M. lateritia* Ab. 3 in HAMPSON 1914a: 65.

rubricincta ab. **ashantica** STRAND, 1916: 82 (*Pseudapiconoma*). Holotype [Ghana] "Ashanti" [BMNH] - an infrasubspecific name for *P. rubricincta* Ab. 1 in HAMPSON 1898: 151.

preussi ab. **brunnea** GRÜNBERG, 1907: 434 (*Pseudapiconoma*). Holotype [Cameroon] „Ngoko-Stat." [ZMHB] - an infrasubspecific name.

burra ab. **congonis** STRAND, 1916: 82 (*Metarctia*). Holotype "Kongo" [BMNH] - an infrasubspecific name for *M. burra* Ab. 1 in HAMPSON 1898: 147.

flavimacula WLK. ab. **elegantissima** STRAND, 1912: 191 (*Pseudapiconoma*). Holotype [Equatorial Guinea] "Nkolentangan" [ZMHB] - an infrasubspecific name.

rufescens ab. **fuscorufescens** STRAND, 1916: 82 (*Metarctia*). Syntype(s) [not presented] [BMNH] - an infrasubspecific name for *M. rufescens* Ab. 1 in HAMPSON 1898: 148.

rubripuncta ab. **hampsoni** STRAND, 1916: 82 (*Metarctia*). Holotype "Congo" [NHRS] - an infrasubspecific name for *M. rubripuncta* Ab. 1 in HAMPSON 1898: 148.

canescens ab. **homoerotica** STRAND, 1916: 82 (*Apisa*). Holotype "Sierra Leone" [BMNH] - an infrasubspecific name for *A. canescens* Ab. 1 in HAMPSON 1898: 143.

elegans n[ew] r[ace] **kivensis** DUFRANE, 1945: 135 (*Balacra*). Holotype [DRC] "Kamituga" [KBIN] - an infrasubspecific name.

lateritia ab. **lateritiola** STRAND, 1916: 82 (*Metarctia*). Syntype(s) "Äthiopisch" [BMNH] - an infrasubspecific name for *M. lateritia* Ab. 1 in HAMPSON 1898: 145.

mariae ab. **latophaga** DUFRANE, 1945: 129 (*Apisa*). Holotype [DRC] „Kamituga" [KBIN] - an infrasubspecific name.

preussi AURIV. ab. *longimaculata* STRAND, 1912: 189-190 (*Balacra*). Holotype [Equatorial Guinea] „Nkolentangan" [ZMHB] - an infrasubspecific name.

testacea ab. *micromacula* STRAND, 1920: 224 (*Balacra*). Syntype(s) "Goldkuste, Nigeria, Gabun, Uganda" [BMNH] - an infrasubspecific name for *B. testacea* Ab. 1. in HAMPSON 1914a: 78.

speculigera ab. *obliterata* GRÜNBERG, 1907: 435 (*Pseudapiconoma*). Holotype [Cameroon] "Buea" [ZMHB] - an infrasubspecific name.

taymansi f.n. *kamitugensis* ab. *obscura* DUFRANE, 1945: 133 (*Metarctia*). Holotype [DRC] "Kamituga" [KBIN] - an infrasubspecific name.

invaria ab. *opobensis* STRAND, 1916: 82 (*Metarctia*). Syntypes [Nigeria] "Opobo in Alt-Calabar" [BMNH] - an infrasubspecific name for *M. invaria* Ab. 2 in HAMPSON 1898: 146.

canescens ab. *perversa* STRAND, 1916: 82 (*Apisa*). Holotype [Kenya] "Sansibar" [BMNH] - an infrasubspecific name for *A. canescens* Ab. 2 in HAMPSON 1898: 143.

rufescens ab. *postfuscescens* STRAND, 1916: 82 (*Metarctia*). Syntype(s) "Äthiopisch" [BMNH] - an infrasubspecific name for *M. rufescens* Ab. 2 in HAMPSON 1898: 148.

preussi ab. *punctata* DUFRANE, 1945: 135 (*Balacra*). Holotype [DRC] „Kamituga" [KBIN] - an infrasubspecific name.

invaria Wlk. cum ab. *pusillima* n. ab. STRAND, 1912: 187-188 (*Metarctia*). Holotype [Cameroon] „Bibundi" [ZMHB] - an infrasubspecific name.

rubicundula STRAND cum ab. *quadrisignatula* n. ab. STRAND, 1912: 187 (*Metarctia*). Holotype [Cameroon] "Mokundange" [ZMHB] - an infrasubspecific name.

rubripuncta Lokalrasse *rosea* AURIVILLIUS, 1905b: 13 (*Metarctia*). Syntypes [DRC] "Mukimbungu" [NHRS] - an infrasubspecific name.

rubricosta KIRIAKOFF, 1957b: 145 (*Metarctia* [*Oenarctia*]) cited by KIRIAKOFF 1960b: 44 as "nom. nov. pro *Metarcia rubripuncta* Hampson, forma *rubricosta* TALBOT, Bull. Hill. Mus., III, p. 125 [1929] [♀]" – the unjustified new name & autorship for *rubricosta* TALBOT, 1929 which under Article 45.6.4. is subspecific name. TALBOTS' statement: "this reddish form with pink hind wing may, perhaps, be only a colour variation" does not mean that the author expressly gave it infrasubspecific rank.

flavimacula ab. *separata* STRAND, 1916: 82 (*Pseudapiconoma*). Syntype(s) "Westafrika" [BMNH] - an infrasubspecific name for the unique specimen of *P. flavimacula* mentioned without any distinguishing character in HAMPSON 1898: 151.

TAXA EXCLUDED FROM THE THYRETINI

The genera and species listed here have been included in the Thyretini in the last catalogue (KIRIAKOFF 1960) or described by KIRIAKOFF (1953a, 1965) as belonging to this tribe. Examination of the type material revealed that they belong to Arctiinae or Syntominae: Syntomini. Valid names are in bold.

Genera

Meganaclia AURIVILLIUS, 1892: 190.

Type species: *Naclia sippia* PLÖTZ, 1880 (by monotypy). Here transferred to Arctiidae: Arctiinae.

Mesonaclia KIRIAKOFF, 1953b: 20, 21.

Type species: *Meganaclia minor* HAMPSON, 1914a (by original designation). A junior synonym of *Nacliodes* STRAND, 1912 (see KIRIAKOFF 1959a).

Nacliodes STRAND, 1912: 183. [proposed as subgeneric name of **Meganaclia**]

Type species: *Meganaclia microsippia* STRAND, 1912 (by monotypy). Here transferred to Arctiidae: Arctiinae.

Pachyceryx KIRIAKOFF, 1957a: 283-284

Type species: *Pachyceryx albomaculata* KIRIAKOFF, 1957 (by original designation). A junior synonym of *Pseudodiptera* KAYE, 1918 (see PRZYBYŁOWICZ & KÜHNE 2008: 147)

Pseudodiptera KAYE, 1918: 229-230.

Type species: *Pseudodipiera musiforme* Kaye, 1918 (by original designation). Transferred to Arctiidae: Syntominae: Syntomini by PRZYBYŁOWICZ & KÜHNE 2008.

Thyrogonia HAMPSON, 1898: 139.

Type species: *Syntomis efulensis* HOLLAND, 1898 (by original designation). Here transferred to Arctiidae: Syntominae: Syntomini.

Species

alberici DUFRANE, 1945: 125-126 (*Ceryx*). Holotype [DRC] "Mohanga" [KBIN]. Transferred to *Pachyceryx* by KIRIAKOFF 1957a. Transferred to Arctiidae: Syntominae: Syntomini by PRZYBYŁOWICZ & KÜHNE 2008. Present taxonomic status: *Pseudodiptera alberici* (DUFRANE, 1945).

albomaculata KIRIAKOFF, 1957a: 284 (*Pachyceryx*). Holotype [DRC] „Tshuapa: Bokuma" [RMCA]. A junior subjective synonym of *Pseudodiptera musiformae* KAYE, 1918 (see PRZYBYŁOWICZ & KÜHNE 2008: 147).

aurantiiventris KIRIAKOFF, 1953a: 96 (*Thyrogonia*). Holotype [DRC] "Tshuapa: Eala" [RMCA]. Here transferred to Syntominae: Syntomini. Present taxonomic status: *Thyrogonia aurantiiventris* KIRIAKOFF, 1953.

clypeatus KIRIAKOFF, 1965: 2 (*Pachyceryx*). Holotype [DRC] "Shabunda (Kivu)" [KBIN]. Transferred to Arctiidae: Syntominae: Syntomini by PRZYBYŁOWICZ & KÜHNE 2008. Present taxonomic status: *Pseudodiptera clypeatus* (KIRIAKOFF, 1965).

cyaneotincta HAMPSON, 1918: 94 (*Thyrogonia*). Holotype [Malawi] "Ruo Valley" [BMNH]. Here transferred to Syntominae: Syntomini. Present taxonomic status: *Thyrogonia cyaneotincta* HAMPSON, 1918

dufranei KIRIAKOFF, 1965: 3 (*Pachyceryx*). Holotype [DRC] "Kabunga (Kivu)" [KBIN]. Transferred to Arctiidae: Syntominae: Syntomini by PRZYBYŁOWICZ & KÜHNE 2008. Present taxonomic status: *Pseudodiptera dufranei* (KIRIAKOFF, 1965).

efulensis HOLLAND, 1898: 12 (*Syntomis*). Holotype [Cameroon] „Efulen, Buele Country" [CMNH]. Transferred to *Thyrogonia* by HAMPSON, 1898. Here transferred to Syntominae: Syntomini. Present taxonomic status: *Thyrogonia efulensis* (HOLLAND, 1898).

hampsoni KIRIAKOFF, 1953a: 95-96 (*Thyrogonia*). Holotype [DRC] "Tshuapa: Eala" [RMCA]. Here transferred to Syntominae: Syntomini. Present taxonomic status: *Thyrogonia hampsoni* KIRIAKOFF, 1953.

microsippia STRAND, 1912: 183 ((*Meganaclia* (*Nacliodes* sg. n.)). Holotype [Equatorial Guinea] "Alen" [ZMHB]. Transferred to *Nacliodes* as separate genus by KIRIAKOFF, 1959. Here transferred to Arctiinae. Present taxonomic status: *Nacliodes microsippia* (STRAND, 1912).

minor HAMPSON, 1914a: 61 (*Meganaclia*). Holotype [Uganda] "Entebbe" [BMNH]. A junior subjective synonym of *Nacliodes microsippia* (STRAND, 1912).

musiforme KAYE, 1918: 230 (*Pseudodiptera*). Holotype [Central African Republic] "Congo, Oubangui-chari, Tschad, Bangui" [BMNH]. Transferred to Arctiidae: Syntominae: Syntomini by PRZYBYŁOWICZ & KÜHNE 2008: 147. Present taxonomic status: *Pseudodiptera musiforme* KAYE, 1918.

sippia PLÖTZ, 1880: 78 (*Naclia*). Syntypes [Cameroon] "Cameroons-Gebirge" [UGD]. Transferred to *Meganaclia* by AURIVILLIUS, 1892. Here transferred to Arctiinae. Present taxonomic status: *Meganaclia sippia* (PLÖTZ, 1880).

trichaetiformis ZERNY, 1912b: 119-120 (*Thyretes*). [Kenya] „Zanzibarkuste" [NHMW]. Here transferred to Syntominae: Syntomini, *incertae sedis*.

Taxonomic changes and comments to the generic level categories

Automolis Hübner, [1819] (1816-[1826]): 170.
Type species: *Sphinx meteus* Stoll, 1781.

Genus *Automolis* is here regarded as a separate from *Metarctia*. My analyses of the morphology of the imago and male genitalia strongly support the separation of these two taxa.

Diagnosis. The main autapomorphies of this genus are the presence of a distinct, sharp process on the scapus; a short but stout tibia terminating with prominent teeth and provided with large epiphisa; a more or less evenly rounded valva (without a distinct costal process); and a wide and short phallus.

Representatives of all species available for study (mostly types) were analyzed for these characters.

Distribution. Currently the genus is restricted to South Africa and (only *A. pallida*) mountainous areas of the Eastern part of the continent (Kenya).

Balacra Walker, 1856: 1721.

Subgenus Balacra Walker, 1856: 1721.
Type species: *Balacra caeruleifascia* Walker, 1856 (by subsequent designation by Kirby 1892: 221).
Subgenus Pseudapiconoma Aurivillius, 1881: 46.
Type species: *Pseudapiconoma testacea* Aurivillius, 1881 (by monotypy).

 Epibalacra Kiriakoff, 1957b: 149. **[syn. nov.]**
 Type species: *Metarctia preussi* Aurivillius, 1904 (by original designation).

According to Kiriakoff (1960) these two subgenera group most of the species of genus *Balacra*. This division is supported by the morphology of the male genitalia. However, in his catalogue *Pseudapiconoma* is treated as a synonym of *Balacra* while *Epibalacra* is a valid subgenus. This statement is based on the wrong interpretation of the male of *B. caeruleifascia*, which was originally described from a female. My examination of numerous specimens belonging to the assemblage of *B. caeruleifascia* and *B. preussi* shows that despite the superficial similarity of both species [especially the females], there are (at least) two characters clearly separating both sexes of these two taxa (see descriptions of species). This discovery enabled the proper ascription of the sexes and reinterpretation of the subgenera.

The males of *B. caeruleifascia* are characterized by a pointed uncus while the males of *B. preussi* have the tip of the uncus flattend. The same flattened type of uncus is shown in the male of *B. testacea* – the type species of *Pseudapiconoma*. Consequently *Epibalacra*, the younger name, is here regarded as a synonym of *Pseudapiconoma*.

The type species of *Pseudapiconoma* is *B. testacea* (see Watson et al. 1980), not *B. flavimacula* as wrongly stated by Kiriakoff (1960: 9).

Subgenus Daphaenisca Kiriakoff, 1953b: 69.
Type species: *Pseudapiconoma daphaena* Hampson, 1898 (by original designation).
 Balacrella Kiriakoff, 1957b: 154. **[syn. nov.]**
 Type species: *Pseudapiconoma affinis* Rothschild, 1910 (by original designation).

Subgenus *Balacrella* is here for the first time synonymized with *Daphaenisca*. Both species belonging here are very similar and a proper identification is possible only after examination of the genital morphology. The differences in genitalia are rather small and do not support the maintenance of two separate subgenera.

Cameroonia PRZYBYŁOWICZ [gen. nov.]

Type species: *Metarctia nigriceps* AURIVILLIUS, 1904 (by present designation)

Diagnosis. Superficially similar to *Mecistorhabdia* and subgenus *Thyretarctia* of *Metarctia* but easily separated by the shape of the male and female antennae. The genitalia are very distinctive. The male has a large, widened, straight phallus. The female has an extremely wide, membranous ductus bursae enlarging gradually to the corpus bursae.

Description. Forewing 16-19 mm. General colouration orange-red with indistinct dark ochraceous markings on forewing; hindwing paler; antenna and cilia ochraceous. Head covered with long, projecting, hair-like scales; labial palpus long, first segment slightly upcurved, second segment 3.5 times as long as wide, third segment small, elongate, 1/3 length of second segment; antenna bipectinate, in male longest rami more than 5 times width of flagellomere, in female rami much shorter, about 2 times width of flagellomere; eye medium-sized, naked; proboscis rudimentary. Tymbal organs absent. Forewing with CuA_1 and CuA_2 parallel to dorsum, 1A+2A convex towards costa in middle. Hindwing with M_2-M_3 separated; Rs and M_1 fused; Sc absent.

Male genitalia: Uncus sclerotized, flattened laterally, pointed apically; tegumen broad; saccus gradually tapering terminally; valva short, wide; costa convex, distal part forming inwardly directed sclerotized process; phallus large, short, stout; vesica provided with two groups of minute cornuti and one large, separate spine.

Female genitalia: Anal papillae relatively large; posterior apophyses twice as long as anterior apophyses; ductus bursae membraneous, almost as wide as corpus bursae, signum present; ductus seminalis slender, entering in proximal part of ductus bursae.

Distribution. Until now known only from Cameroon.

Etymology. The generic name is derived from the name of the country [Cameroon] and refers to the known distribution of the single known species. It is feminine in gender.

Remarks. The species was originally described in genus *Metarctia*. It was ascribed to the subgenus *Thyretarctia* by KIRIAKOFF (1957c: 99) based on superficial similarity and postulated congruence of the morphology of the male genitalia with those of *M.* (*T.*) *schoutedeni*. The reexamination of the genitalia clearly shows that both the morphology of the valva (shape and position of distal processes) and phallus do not match the general pattern characteristic not only for *Thyretarctia* but also for the whole genus *Metarctia*.

Metarctia WALKER, 1855: 769.

Type species: *Metarctia rufescens* WALKER, 1855 (by monotypy).

Oenarctia KIRIAKOFF, 1953b: 51. [syn. nov.]

Type species: *Oenarctia erlangeri* ROTHSCHILD, 1910 (by original designation).

Oenarctia is here for the first time proposed as a synonym of *Metarctia*. The morphology of the male genitalia (especially the shape of the valva) shows no significant differences from that of the type species and other members of the nominotypical subgenus. *Metarctia* s. str. is a large assemblage of species in which several types of male genitalia can be found. Members of the former *Oenarctia* can be classified as one of such groups. However some species of *Metarctia* s. str. show intermediate stages between groups so maintaining subgenus *Oenarctia* is unjustified.

A similar situation probably concerns subgenus *Pinheyata*. However, the lack of material available for study does not allow for the formal synonymization of the subgenus.

Rhipidarctia KIRIAKOFF, 1953b: 28.

Subgenus *Rhipidarctia* (s. str.)

Type species: *Elsa rubrosuffusa* KIRIAKOFF, 1953b (by original designation).

      Elsita KIRIAKOFF, 1954c: 29 (nom. nov. pro *Elsa* KIRIAKOFF, 1953)   **[syn. nov.]**

      Elsa KIRIAKOFF, 1953b: 23 [nec HONRATH, 1892].

      Type species: *Anace invaria* WALKER, 1856 (by original designation).

      Takwa KIRIAKOFF, 1957c: 96.   **[syn. nov.]**

      Type species: *Takwa xenops* KIRIAKOFF, 1957 (by original designation).

Monotypical genus *Takwa* KIRIAKOFF, 1957 is here for the first time synonymized with the nominotypical subgenus of the genus *Rhipidarctia* KIRIAKOFF, 1953. Reexamination of the holotype of *T. xenops* KIRIAKOFF, 1957 (the type species of genus *Takwa*) indicated that it is the typical representative of the genus *Rhipidarctia*. The morphology of the male genitalia locates it in subgenus *Rhipidarctia*. Very narrow forewing regarded by KIRIAKOFF as character clearly differing both genera can not be treted as such. Tendencion for narrowing of the forewing is observed in most of the members of the subgenus *Rhipidarctia*.

Subgenus *Elsita* KIRIAKOFF, 1954 of genus *Rhipidarctia* KIRIAKOFF, 1953 is here for the first time synonymized with the nominotypical subgenus of the genus. Both of these subgenera were proposed by KIRIAKOFF to separate groups of species differing mostly in the shape of the valva in the male genital organs. *Anace invaria* WALKER was selected by him as type species of this new subgenus. However, *A. invaria* was described from a female and in designating subgenus *Elsita*, KIRIAKOFF wrongly interpreted a male of another species belonging to this new subgenus as a male of *R. invaria*. This mistake is here corrected.

The genital organ of the truth male of *R. invaria* is typical for the members of subgenus *Rhipidarctia*. As a result of this mistake *Elsita* KIRIAKOFF, 1954 becomes a junior objective synonym of *Rhipidarctia* KIRIAKOFF, 1953. The species formerly referred to subgenus *Elsita* require a new name, as follows.

New name:

Genus: *Rhipidarctia* KIRIAKOFF, 1953b: 28.

Subgenus: *Elsitia* PRZYBYŁOWICZ [nom. nov.]

Type species: *Metarctia pareclecta* HOLLAND, 1893: 395-396.

DIAGNOSIS. Adults generally larger than representatives of the other subgenera; flagellomeres of males bipectinate with long rami; forewing always with dark suffusion in middle of interspace between M2 and A1-A2 (in rare specimens this suffusion is entirely missing); pseudovalva narrow, not widened terminally.

Family: Arctiidae

Subfamily: Syntominae

Tribe: *Thyretini*

Apisa WALKER, 1855

Subgenus: *Apisa* WALKER, 1855

1. *Apisa (Apisa) arabica* WARNECKE, 1934
(Pl. I)

Short diagnosis. Imago. Forewing length: 22 mm. No external differences from *A. canescens* have been found.

Male genitalia. Not examined.

Female genitalia. Unknown.

Early stages. Unknown.

Biology. Unknown except for the collecting dates (June, August, September).

Distribution. Oman, Saudi Arabia, Yemen.

Remarks. The taxonomic status of this species with respect to *A. canescens* remains uncertain. HACKER (1999) and WILTSHIRE (1980a, 1980b, 1990) treat it as a subspecies of *A. canescens*. The species was described from an unstated number of specimens. Despite an extensive search only one syntype has been located in the Lepidoptera collection of ZIMH: a male collected in San'aa, 22.6.31. The specimen lacks the abdomen, but a genitalia slide was not found (pers. comm. H. RIEFENSTAHL). According to WEIDNER (1974) remaining syntypes have been destroyed during the Second World War. In BMNH another male may possibly belong to the type series. It bears a label with the handwritten word "paratype", but the collecting date (20.VIII.1931) is different than those given in the original description. It can be also the worn specimen mentioned by WARNECKE: „♂ und ♀ von San'a, 22.6.31, 9.9.31, 16.9.31, bis auf 1 ♂ mehr oder minder stark beschädigt". The specimen is almost without scales on the wings, but otherwise very similar to *A. canescens*.

The holotype female of *A. canescens lippensi*, here regarded as a synonym of *A. arabica*, has not been located in KBIN. *Apisa arabica* was omitted in KIRIAKOFF's catalogue (1960) as well as in his description of *A. canescens lippensi* in which the holotype (♀) is described as the first specimen belonging to "famille Thyretidae" collected outside Africa.

Extensive, new material from both sides of the Red Sea is required to compare the morphological variability of the populations.

---

2. *Apisa (Apisa) canescens* WALKER, 1855
(Pl. I; ♂ Pl. 1; ♀ Pl. 19)

Short diagnosis. Imago. Forewing length: 13-26 mm. Very variable in shape and colouration. Whole body more or less uniformly creamy white to grey

but never silver, brown, or with distinctively darker costal area.

Male genitalia. Harpa very long, at least as long as valva.

Female genitalia. Ductus bursae narrow, elongate; signum narrowed medially, moderately sclerotized, situated medially.

37

Early stages. Caterpillar creamy white with rows of long hair in tufts (PINHEY 1975).

Biology. Prefers dry, open areas. The caterpillar feeds on *Cosmos* sp. (Asteraceae) (SEVASTOPULO 1975). Adults collected throughout the year.

Distribution. Burundi, Congo, DRC, Gabon, Kenya, Namibia, Nigeria, RSA, Senegal, Sierra Leone, Somalia, Tanzania, Zimbabwe.

Remarks. The most widely distributed of all Thyretini. The taxonomic status of the different populations, forms, and species described especially from the northern part of the range (Libya, Saudi Arabia) require further investigation. The subspecific variability is not well known.

Four syntypes without indication of their sex are mentioned in the original description. Two of them (♂♂) have been found in the BMNH. One specimen, with a small rounded label "TYPE" was selected and here designated as the LECTOTYPE to prevent any doubts as to the identity of this species. The second specimen was labeled as the PARALECTOTYPE.

Subgenus *Dufraneella* KIRIAKOFF, 1953

3. *Apisa (Dufraneella) fontainei* KIRIAKOFF, 1959
(Pl. I; ♂ Pl. 1)

Short diagnosis. Imago. Forewing length: 14 mm. Pale, greyish white with distinctly darker costa; similar to *A. subcanescens* but flagellum dark, concolorous with costa.

Male genitalia. Indistinguishable from *A. subcanescens*.

Female genitalia. Unknown.

Early stages. Unknown.

Biology. Unknown except for the collecting dates (April). In Kakamega Forest (Kenya) moths have been collected at an artificial light source located inside the secondary forest.

Distribution. DRC, Kenya, Rwanda.

Remarks. This species is very similar to *A. subcanescens*. Both of them are known from a few specimens and more extensive material is needed to clarify the status of these two taxa. The male genitalia of the holotype are very worn and difficult to interpret (see PRZYBYŁOWICZ & KÜHNE 2008).

4. *Apisa (Dufraneella) grisescens* (DUFRANE, 1945)
(Pl. I; ♂ Pl. 1)

Short diagnosis. Imago. Forewing length: 11-13 mm. Small, dark species similar to *A. hildae* and *A. rendalli*. Differs from the first by the opaque forewing, from the last by the distinctly darker costal area of the forewing.

Male genitalia. Harpa reduced to small teeth in basal part of valva.

Female genitalia. Unknown.

Early stages. Unknown.

Biology. Unknown except for the collecting dates (April, December).

Distribution. DRC, Malawi, Tanzania.

5. *Apisa (Dufraneella) hildae* KIRIAKOFF, 1961
(Pl. I; ♂ Pl. 1)

Short diagnosis. Imago. Forewing length: 12 mm. Habitus and size as in *A. rendalli* but wings semi-transparent except for costal area of forewing.

Male genitalia. Vesica with sclerotized tooth in basal part, distal margin of valva only slightly concave.

Female genitalia. Unknown.

Early stages. Unknown.

Biology. Unknown except for the collecting dates (March, April).

Distribution. Namibia.

6. *Apisa (Dufraneella) rendalli* ROTHSCHILD, 1910
(Pl. I; ♂ Pl. 1)

Short diagnosis. Imago. Forewing length: 12-14 mm. One of the smallest species. Characterized by the dark brown, uniform colouration of the body.

Male genitalia. Saccus elongate; distal margin of valva deeply concave.

Female genitalia. Unknown.

Early stages. Unknown.

Biology. Unknown except for the collecting dates (March, December).

Distribution. DRC, Malawi.

Remarks. All three syntypes (♂♂) have been found in the collection of BMNH. One of them, with a small rounded label "TYPE" was selected and here designated as LECTOTYPE to prevent any doubts as to the identity of this species. Remaining two males were labeled as the PARALECTOTYPES.

7. *Apisa (Dufraneella) subcanescens* ROTHSCHILD, 1910
(Pl. I; ♂ Pl. 1)

Short diagnosis. Imago. Forewing length: 15-22 mm. Pale, brownish grey species with semi-transparent wings; forewing slightly pointed toward apex with intensive dark brown suffusion along costa. Superficially very similar to *A. cinereocostata*. From *A. fontainei* it differs by the pale, brownish white flagellum.

Male genitalia. Indistinguishable from those of *A. fontainei*.

Female genitalia. Not examined.

Early stages. Unknown.

Biology. Unknown.

Distribution. Senegal.

Remarks. The species was described from unstated number of specimens. Two syntypes (male and female) have been found in the BMNH. The male was selected and here designated as LECTOTYPE. This specimen bears a small, rounded, red-bordered "TYPE" label. The female was labeled as the PARALECTOTYPE.

Subgenus *Parapisa* Kiriakoff, 1952

8. *Apisa (Parapisa) cinereocostata* Holland, 1893
(Pl. I; ♂ Pl. 1)

Short diagnosis. Imago. Forewing length: 12-14 mm. Pale grey species characterized by slightly more brownish, not greyish, colouration of the forewing. Proper determination possible only after examination of male genitalia.

Male genitalia. Uncus apically with two widely separated hook-like processes.

Female genitalia. Unknown.

Early stages. Unknown.

Biology. Unknown except for the collecting dates (January, May, August, October).

Distribution. Gabon, Guinea, Ivory Coast, Nigeria.

Remarks. The species was confused with *A. canescens* for a long time but it differs in the shape of the male genitalia.

9. *Apisa (Parapisa) subargentea* Joicey & Talbot, 1921
(Pl. I; ♂ Pl. 1; ♀ Pl. 19)

Short diagnosis. Imago. Forewing length: 14-16 mm. Differs from all other *Apisa* species by the shiny, silver colouration of the forewing.

Male genitalia. Uncus apically with two narrow, parallel lobes.

Female genitalia. Ductus bursae short and wide; signum single, sclerotized, situated in proximal part of corpus bursae.

Early stages. Unknown.

Biology. Unknown except for the collecting dates (March, May, August, September, November, December). In Kakamega Forest (Kenya) moths have been collected only at an artificial light source located in open habitat (farmland).

Distribution. Burundi, DRC, Kenya, Rwanda.

Remarks. See Przybyłowicz & Kühne (2008).

Apisa INCERTAE SEDIS

10. *Apisa manettii* Turati, 1924

Short diagnosis. Imago. Forewing length: 17-21 mm. Original description insufficient for proper determination.

Male genitalia. Unknown, not presented by Turati (1924).

Female genitalia. Unknown, not presented by Turati (1924).

Early stages. Unknown.

Biology. Unknown except for the collecting dates (August, October).

Distribution. Libya (Cyrenaica).

Remarks. Type not located, probably in MRSN. Possibly this taxon belongs to *A. canescens* but additional material is needed for detailed examination.

Automolis Hübner, [1819] 1816

11. *Automolis bicolora* (Walker, 1856)
(Pl. I; ♂ Pl. 1)

Short diagnosis. Imago. Forewing length: 12-15 mm. Similar to *A. meteus* but smaller, with blackish mesothorax between tegulae and reduced yellow colouration of costa and base of forewing; wings (especially hindwing) semi-transparent.

Male genitalia. Very similar to those of *A. meteus* and *A. incensa* but sacculus almost straight, only slightly convex.

Female genitalia. Unknown.

Early stages. Unknown.

Biology. Unknown except for the collecting dates (January, February, December).

Distribution. RSA.

Remarks. Female unknown but probably wingless like other known representatives of the genus.

12. *Automolis crassa* (Felder, 1874)
(Pl. I; ♂ Pl. 2; ♀ Pl. 19)

Short diagnosis. Imago. Forewing length: 13-18 mm. Small species, entirely pale brown.

Male genitalia. Distal process of valva narrow, elongate.

Female genitalia. Ventral pheromone glands pouch shaped, without elongate, finger-like processes.

Early stages. Unknown.

Biology. Unknown except for the collecting dates (August, October).

Distribution. RSA.

13. *Automolis incensa* (Walker, 1864)
(Pl. I; ♂ Pl. 2)

Short diagnosis. Imago. Forewing length: 21-25 mm. The largest representative of the genus. Head, thorax, abdomen, base of wings and cilia yellow, remaining parts of wings dark brown. Female indistinguishable from that of *A. meteus*, but slightly larger.

Male genitalia. Indistinguishable from those of *A. meteus*.

Female genitalia. Unknown.

Early stages. Unknown.

Biology. Unknown except for the collecting dates (October, November).

Distribution. RSA.

14. *Automolis meteus* (Stoll, 1780-82)
(Pl. I; ♂ Pl. 2; ♀ Pl. 19)

Short diagnosis. Imago. Forewing length: 16-19 mm. Very characteristic yellow species with brownish-black antennae and wings except for costa, base of forewing and cilia of both pairs of wings. Female yellow, unicolorous, wingless.

Male genitalia. Indistinguishable from those of *A. incensa*.

Female genitalia. Ventral pheromone glands with elongate, finger-like processes terminally.

Early stages. Caterpillar and pupa described and illustrated by FAWCETT (1903).

Biology. The caterpillar feeds on *Gnidia, Lasiosiphon, Acalypha* (FAWCETT 1903), *Trema bracteolata* BLUME (JANSE 1945) and different species of grasses (TAYLOR 1949). According to the last author the larvae are gregarious, active at night, and hiding during the day in holes and burrows in the ground. There are two generations each the year. Unidentified species of Tachinidae and Ichneumonidae have been reared from the caterpillars. Adults collected in January-May, September, October, December.

Distribution. RSA, Swaziland.

15. *Automolis pallida* (HAMPSON, 1901)
(Pl. I; ♂ Pl. 2)

Short diagnosis. Imago. Forewing length: 13-17 mm. General colouration pale brown; head and thorax dark brown; hindwing pale, almost creamy, semi-transparent; base of forewing paler than distal part.

Male genitalia. Valva elongate, approximately two times longer then wide.

Female genitalia. Unknown.

Early stages. Unknown.

Biology. Unknown except for the collecting dates (February, March).

Distribution. Kenya.

Remarks. This is the only species of the genus that is absent in South Africa and restricted in range to the mountainous regions of Eastern Africa.

Balacra WALKER, 1856

Subgenus *Balacra* WALKER, 1856

16. *Balacra (Balacra) belga* KIRIAKOFF, 1954
(Pl. I)

Short diagnosis. Imago. Forewing length: 22 mm. Very similar to *B. nigripennis*. Can be separated by the presence of white markings on the hindwing and white, not transparent, markings on the forewing.

Male genitalia. Unknown.

Female genitalia. Not examined.

Early stages. Unknown.

Biology. Unknown except for the collecting date (September).

Distribution. DRC.

Remarks. So far only one female is known. It is most probably an individual colour form of *B. nigripennis*.

17. *Balacra (Balacra) caeruleifascia* WALKER, 1856
(Pl. I; ♂ Pl. 2; ♀ Pl. 20)

Short diagnosis. Imago. Forewing length: 20-30 mm. The only species of the subgenus with a reddish rust thorax and abdomen; markings on forewing paler than ground colour. Female similar to that of *B. (P.) preussi* but without transparent blotch between $CuA_1 - CuA_2$ and without grey narrow band along distal margin of tergites.

Male genitalia. Indistinguishable from those of *B. guillemei*. Uncus pointed as in remaining representatives of the subgenus.

Female genitalia. Signum sclerotized, longer and narrower than in *B. nigripennis*.

Early stages. Unknown.

Biology. The caterpillar was found in Uganda feeding on "dry scabs an stems and dry leaves of *Coffea* sp." (LE PELLEY 1959). Adults collected in January-April, June-August, December.

Distribution. Cameroon, CAR, Congo, DRC, Equatorial Guinea, Gabon, Ghana, Ivory Coast, Liberia, Nigeria, Sierra Leone, Uganda.

Remarks. The record from Uganda is cited from biological data published by LE PELLEY (1959). No specimens were found in collections.

18. *Balacra (Balacra) guillemei* (OBERTHÜR, 1911)
(Pl. I; ♂ Pl. 2)

Short diagnosis. Imago. Forewing length: 19-25 mm. Similar to *B. rattrayi* but forewing more suffused with reddish scales; blotches transparent; thorax creamy white with distinctive red strikes on tegulae, between them, and on distal margin of patagia. Female unknown.

Male genitalia. Indistinguishable from those of *B. caeruleifascia.*.

Female genitalia. Unknown.

Early stages. Unknown.

Biology. Unknown except for the collecting dates (March, April, November-December).

Distribution. DRC.

19. *Balacra (Balacra) nigripennis* (AURIVILLIUS, 1904)
(Pl. I; ♂ Pl. 2; ♀ Pl. 20)

Short diagnosis. Imago. Forewing length: 23-30 mm. Very distinctive black species with hyaline or rarely (males) white markings on forewing; hindwing entirely black.

Male genitalia. Vesica provided at base with large sclerotized process.

Female genitalia. Ductus bursae and corpus bursae membranous; signum large, elongate, sclerotized.

Early stages. Unknown.

Biology. Unknown except for collecting dates (January-May, August-December).

Distribution. Angola, DRC.

Remarks. *Balacra belga* is probably a colour form of this species.

20. *Balacra (Balacra) rattrayi* (Rothschild, 1910)
(Pl. I; ♂ Pl. 2; ♀ Pl. 20)

Short diagnosis. Imago. Forewing length: 20-27 mm. Male differing by pale olive grey forewing with fascia of yellow blotches in middle part and two short, red streaks near base. Female forewing similarly coloured but darker, entirely olive grey and with blotches transparent.

Male genitalia. Terminal part of valva widely rounded.

Female genitalia. Signum reduced, narrow, elongate, weakly sclerotized.

Early stages. A short description of the caterpillar and pupa has been provided by Fontaine (1992).

Biology. Caterpillars fed in captivity on *Pelargonium* (Geraniaceae); pupation lasted 28-29 days (Fontaine 1992). According to Le Pelley (1959), another host plant (in Uganda) is *Gossypium* sp. Adults collected in April, May, September-December. In Kakamega Forest (Kenya) moths have been collected at an artificial light source located inside the middle-aged secondary forest.

Distribution. Burundi, DRC, Kenya, Rwanda, Uganda.

Subgenus *Callobalacra* Kiriakoff, 1953

21. *Balacra (Callobalacra) alberici* Dufrane, 1945
(Pl. I; ♂ Pl. 2; ♀ Pl. 20)

Short diagnosis. Imago. Forewing length: 19-22 mm. Differs from the other *Callobalacra* species by its uniformly pale brown forewing.

Male genitalia. Distal process of valva large, flat; phallus short, wide; vesica with numerous cornuti.

Female genitalia. Colliculum short.

Early stages. Unknown.

Biology. Unknown except for the collecting dates (February-May, July, October, December).

Distribution. DRC.

Remarks. The species was synonymized with *B. jaensis* by Kiriakoff (1957b). The two species are almost identical in male genitalia but the differences in colouration and distribution suggest that they are distinct. Comparing the genitalia of longer series may help in the proper interpretation of these two taxa. Only 21 males and 1 female are known until now.

22. *Balacra (Callobalacra) jaensis* Bethune-Baker, 1927
(Pl. I; ♂ Pl. 3)

Short diagnosis. Imago. Forewing length: 18-23 mm. Similar to *B. (C.) rubrostriata* but smaller and with forewing paler with less contrasting markings; abdomen without red patches.

Male genitalia. Almost identical to those of *B. alberici*; phallus slightly longer, narrower.

Female genitalia. Not examined.

Early stages. Unknown.

Biology. Unknown except for the collecting dates (April, October, November).

Distribution. Cameroon, Gabon, Nigeria.

Remarks. So far only seven males and one female are known. To prevent the loss of the important colour pattern on the abdomen of the female, it was not dissected.

23. *Balacra (Callobalacra) rubrostriata* (AURIVILLIUS, 1898)
(Pl. I; ♂ Pl. 3; ♀ Pl. 21)

Short diagnosis. Imago. Forewing length: 23-33 mm. *Balacra rubrostriata* is easily distinguished from all other Thyretini by the following combination of characters: large size, whitish yellow forewing with dark veins, and red patches on both sides of each abdominal segment.

Male genitalia. Distal process of valva small, narrow; phallus long; vesica provided with cornuti only at base.

Female genitalia. Colliculum long.

Early stages. Unknown.

Biology. Unknown. Adults collected throughout the year. In Kakamega Forest (Kenya) moths have been collected at an artificial light source located both inside the middle-aged and young secondary forest and in open habitats (farmland, grassland).

Distribution. Burundi, Cameroon, DRC, Gabon, Ghana, Kenya, Togo, Uganda, Zambia.

Subgenus *Compsochromia* KIRIAKOFF, 1953

24. *Balacra (Compsochromia) compsa* (JORDAN, 1904)
(Pl. I; ♂ Pl. 3; ♀ Pl. 21)

Short diagnosis. Imago. Forewing length: 19-25 mm. Male similar to next species (for differences, see below). Female entirely brownich black above except for hyaline markings on wings.

Male genitalia. Uncus wide; valva terminated by prominent spine; base of vesica with large, sclerotized, and protruding projection.

Female genitalia. Ostium bursae small, irregular, with few plicae convergent towards middle; ductus bursae slender, as long as corpus bursae; signum elongate, divided transversaly into several parts; ventral pheromone glands sack-like.

Early stages. Unknown.

Biology. Unknown except for the collecting dates (January-May, July, September-November). In Kakamega Forest (Kenya) moths have been collected at an artificial light source located inside the middle-aged secondary forest.

Distribution. Angola, Burundi, DRC, Kenya, Rwanda, Uganda.

25. *Balacra (Compsochromia) diaphana* KIRIAKOFF, 1957
(Pl. I; ♂ Pl. 3)

Short diagnosis. Imago. Forewing length: 18-25 mm. Resembling *B. (C.) compsa*. Male with reduced, dark colouration on forewing: discal cell hyaline with minute black dot in middle, dark patch in basal part of wing not reaching vein CuA_1, interspace between veins M_3 and CuA_1 with basal part black (in *compsa*, basal part hyaline). Female differing by much reduced hyaline spots on forewing and presence of lateral, red patches above base of hindwing and on first abdominal segment.

Male genitalia. Uncus narrow, valva terminated with small spine, base of vesica with small, sclerotized projection.

Female genitalia. Not examined.

Early stages. Unknown.

Biology. Unknown except for the collecting dates (September-December).

Distribution. CAR, Kenya, Uganda.

Remarks. To prevent the loss of the important colour pattern of the female abdomen, the only known female specimen was not dissected.

Subgenus *Daphaenisca* Kɪʀɪᴀᴋᴏꜰꜰ, 1953

26. *Balacra (Daphaenisca) affinis* (Rᴏᴛʜsᴄʜɪʟᴅ, 1910)
(Pl. I; ♂ Pl. 3; ♀ Pl. 21)

Short diagnosis. Imago. Forewing length: 16-21 mm. Wings, thorax and abdomen dark ochraceous; vertex, antennae, patagia, legs and small patches on thorax and basal segments of abdomen yellow; distal margin of each tergite red. Both species in this subgenus are very similar. For proper determination an examination of the male genitalia is required.

Male genitalia. Uncus trifurcate terminally, mediodorsal process often small; saccus narrow.

Female genitalia. Ductus bursae elongate, slender, sclerotized in proximal part, then membranous.

Early stages. Unknown.

Biology. Unknown. Adults collected throughout the year.

Distribution. Cameroon, CAR, Congo, DRC, Gabon, Uganda.

27. *Balacra (Daphaenisca) daphaena* (Hᴀᴍᴘsᴏɴ, 1898)
(Pl. II; ♂ Pl. 3; ♀ Pl. 21)

Short diagnosis. Imago. Forewing length: 16-20 mm. Indistinguishable from *B. affinis*.

Male genitalia. Uncus bifurcate terminally; saccus wide.

Female genitalia. Ductus bursae short, sclerotized plate at 1/3 of length.

Early stages. Unknown.

Biology. Unknown except for the collecting dates (January, April, October-December).

Distribution. Cameroon, Nigeria.

Subgenus *Heronina* Kɪʀɪᴀᴋᴏꜰꜰ, 1955

28. *Balacra (Heronina) herona* (Dʀᴜᴄᴇ, 1887)
(Pl. II; ♂ Pl. 3; ♀ Pl. 22)

Short diagnosis. Imago. Forewing length: 21-32 mm. Similar to *Balacra (L.) elegans* but larger, stouter built, upper part of thorax pale creamy brown with two pairs of parallel, narrow, red lines: one on tegulae and other between tegulae.

Male genitalia. Uncus extremely wide and short, with four processes: outer processes sharp, elongate; inner processes short, obtuse.

Female genitalia. Ostium bursae large, oval, with plicae convergent towards middle.

Early stages. Unknown.

Biology. Unknown except for the collecting dates (February-April, June-December).

Distribution. Cameroon, CAR, DRC, Equatorial Guinea, Gabon, Ghana, Ivory Coast, Nigeria, Sierra Leone, Chad, Uganda.

Subgenus *Lamprobalacra* KIRIAKOFF, 1953

29. *Balacra (Lamprobalacra) elegans* AURIVILLIUS, 1892
(Pl. II; ♂ Pl. 3; ♀ Pl. 22)

Short diagnosis. Imago. Forewing length: 17-26 mm. May be confused with *Balacra (H.) herona* but without red, narrow lines on mesothorax and tegulae. Colouration of patagia also much paler than rest of upper part of thorax. Female with forewing olive and hindwing dark yellow.

Male genitalia. Sclerotized process of phallus very small.

Female genitalia. Ostium bursae very small with few convergent plicae.

Early stages. Unknown.

Biology. Unknown. Adults collected throughout the year.

Distribution. Cameroon, DRC, Equatorial Guinea, Gabon, Ghana, Ivory Coast, Nigeria, Uganda.

30. *Balacra (Lamprobalacra) furva* HAMPSON, 1911
(Pl. II; ♂ Pl. 3)

Short diagnosis. Imago. Forewing length: 16-19 mm. Head red with creamy stripe between antennae. Thorax and abdomen with cream and red pattern. Hindwing red with cream cilia. Similar to *B. rubricincta* but colouration of forewing darker, olive brown, and transparent middle part usually more prominent.

Male genitalia. As *B. rubricincta* but sclerotized process of phallus longer and narrower, and base of costa of valva only slightly convex.

Female genitalia. Unknown.

Early stages. Unknown.

Biology. Unknown except for the collecting dates (January, February, May, August-November).

Distribution. Cameroon, Ghana, Ivory Coast.

Remarks. The female remains unknown.

31. *Balacra (Lamprobalacra) pulchra* AURIVILLIUS, 1892
(Pl. II; ♂ Pl. 4; ♀ Pl. 22)

Short diagnosis. Imago. Forewing length: 19-31 mm. Very characteristic and unique in colouration: almost entirely white except for partly red legs and head. Forewing semi-transparent in middle.

Male genitalia. Sclerotized process of phallus very large, angular, without membranous tip.

Female genitalia. Distal margin of 8th sternite not deeply concave, lateral lobes reaching only slightly behind ostium bursae.

47

Early stages. Unknown.

Biology. Adults collected throughout the year. In DRC imagos have been observed to be active in the first hours of the night and have been attracted to the light when *B. flavimacula* just stopped to be active (J. & W. DE PRINS, pers. comm.). In Kakamega Forest (Kenya) moths have been collected at an artificial light source located inside the middle-aged and young secondary forest.

Distribution. Angola, Cameroon, CAR, Chad, DRC, Equatorial Guinea, Gabon, Ghana, Ivory Coast, Kenya, Nigeria, Uganda.

Remarks. Very common in Central Africa.

32. *Balacra (Lamprobalacra) rubricincta* HOLLAND, 1893
(Pl. II; ♂ Pl. 4; ♀ Pl. 22)

Short diagnosis. Imago. Forewing length: 18-24 mm. Similar to *B. furva* in red-white pattern on upper part of thorax and abdomen, but can be separated by olive-yellow forewing with slightly semi-transparent middle part. Female much bigger with pinkish-olive, not transparent forewing.

Male genitalia. Sclerotized process of phallus short and wide, base of costa of valva distinctly convex.

Female genitalia. Distal margin of 8th sternite deeply concave, lateral lobes reaching well beyond ostium bursae.

Early stages. Unknown.

Biology. Unknown. Adults collected throughout the year. In Kakamega Forest (Kenya) moths have been collected at an artificial light source located both inside the middle-aged and young secondary forest and in open habitat (grassland).

Distribution. Cameroon, DRC, Equatorial Guinea, Ghana, Ivory Coast, Kenya, Nigeria, Uganda.

Remarks. The proper ascription of females to *B. rubricincta* or *B. furva* is questionable and requires further collecting or rearing.

Subgenus *Pseudapiconoma* AURIVILLIUS, 1881

33. *Balacra (Pseudapiconoma) basilewskyi* KIRIAKOFF, 1953
(Pl. II; ♂ Pl. 4)

Short diagnosis. Imago. Forewing length: 21-22 mm. Distinctive species with bright red forewing provided with few orange-yellow blotches, and contrasting yellow hindwing.

Male genitalia. Valva tipped with sharp, sclerotized spine.

Female genitalia. Unknown.

Early stages. Unknown.

Biology. Unknown except for the collecting dates (July, October, November).

Distribution. Cameroon, DRC.

Remarks. The female remains unknown.

34. *Balacra (Pseudapiconoma) batesi* (Druce, 1910)
(Pl. II; ♂ Pl. 4; ♀ Pl. 23)

Short diagnosis. Imago. Forewing length: 19-26 mm. Similar to *B. flavimacula* in brown forewing with yellow blotches with narrow, red rim. It can be at once separated by the presence of a distinct yellow blotch between Cu_1 and Cu_2.

Male genitalia. Basal part of uncus very narrow.

Female genitalia. Dorsal and ventral pheromone glands without digitate processes.

Early stages. Unknown.

Biology. Unknown except for the collecting dates (January – March, June-October).

Distribution. Cameroon, DRC, Equatorial Guinea, Gabon, Ghana, Ivory Coast, Nigeria, Uganda.

35. *Balacra (Pseudapiconoma) flavimacula* Walker, 1856
(Pl. II; ♂ Pl. 4; ♀ Pl. 23)

Short diagnosis. Imago. Forewing length: 16-23 mm. Very variable in colouration and degree of reduction of yellow markings on forewing. Similar to *B. batesi* and *B. monotonia*. Differs from the first by the absence of a yellow or red blotch between Cu_1-Cu_2 (some individuals may have a trace of red dot). Yellow markings in *B. monotonia* greatly reduced with only one or two small red blotches with yellow center.

Male genitalia. Indistinguishable from *B. monotonia* and *B. preussi.*

Female genitalia. Dorsal and ventral pheromone glands with short digitate processes. Indistinguishable from *B. monotonia.*

Early stages. The caterpillar and pupa have been described by Alibert (1951).

Biology. Caterpillar found on cacao (*Theobroma cacao*) (Alibert 1951), *Carica* sp., *Sonchus* sp. (Sevastopulo 1975), and in Uganda on *Coffea* sp., *Morus* sp. and *Solanum* sp. (le Pelley 1959). Adults collected in January, March, April, June, August-December. In DRC imagos have been observed to be active and attracted to the light exclusively shortly after sunset (J. & W. de Prins, pers. comm.). In Kakamega Forest (Kenya) moths have been collected at an artificial light source located inside the young secondary forest.

Distribution. Angola, DRC, Gabon, Ghana, Guinea, Ivory Coast, Kenya, Nigeria, Tanzania, Uganda.

Remarks. *Balacra flavimacula* is widespread and often very common in tropical Africa. Several colour forms without taxonomic value have been described. The relationships with *B. haemalea* and *B. monotonia* are not clear and further research is badly needed to clarify the taxonomy of this group of taxa.

The record form Uganda is cited from biological data published by le Pelley 1959. No specimens were found in collections.

For further taxonomic remarks see Przybyłowicz & Kühne 2008: 148.

36. *Balacra (Pseudapiconoma) fontainei* Kiriakoff, 1953
(Pl. II)

Short diagnosis. Imago. Forewing length: 20-22 mm. Differs from all other species in the subgenus by the orange body, forewing suffused with brown scales and provided with a series of orange yellow blotches in its median part and with one in the cell.

Male genitalia. Not examined.

Female genitalia. Unknown.

Early stages. Unknown.

Biology. Unknown except for the collecting dates (January, September).

Distribution. DRC.

Remarks. The species is known only from males. Only the holotype was examined. Because of the lack of additional material and because the species is easy to separate based on its colouration, the genitalia were not analyzed.

37. *Balacra (Pseudapiconoma) haemalea* HOLLAND, 1893
(Pl. II)

Short diagnosis. Imago. Forewing length: 20-24mm. Similar to *B. basilewskyi* but entire body red. Basal half of forewing suffused with blackish grey, shiny scales. Markings on forewing red, paler than ground colouration.

Male genitalia. Not examined.

Female genitalia. Unknown.

Early stages. Unknown.

Biology. Unknown except for the collecting dates (April, August-November).

Distribution. Cameroon, DRC, Equatorial Guinea, Gabon.

Remarks. Possibly only an individual colour form of *B. basilewskyi* or *B. flavimacula*.

38. *Balacra (Pseudapiconoma) humphreyi* ROTHSCHILD, 1912
(Pl. II; ♂ Pl. 4)

Short diagnosis. Imago. Forewing length: 19-23 mm. Forewing uniformly brown except for distinctive yellow blotch between CuA-A1A2 and additional small dots in cell and between M_1-M_3.

Male genitalia. Phallus almost straight.

Female genitalia. Not examined.

Early stages. Unknown.

Biology. Unknown except for the collecting dates (February).

Distribution. Ghana, Ivory Coast, Nigeria, Sierra Leone.

Remarks. Dot between M_1-M_2 sometimes absent.

39. *Balacra (Pseudapiconoma) monotonia* (STRAND, 1912)
(Pl. II; ♂ Pl. 4; ♀ Pl. 23)

Short diagnosis. Imago. Forewing length: 18-25 mm. The only species with an entirely brown forewing, exceptionally with a trace of yellow-red dots.

Male genitalia. Indistinguishable from those of *B. flavimacula* and *B. preussi*.

Female genitalia. Indistinguishable from those of *B. flavimacula*.

Early stages. Unknown.

Biology. Unknown except for the collecting dates (April, October, November).

Distribution. Angola, Cameroon, Congo, DRC, Equatorial Guinea.

Remarks. Taxonomic status unclear. It could be a colour form of *B. flavimacula*.

40. *Balacra (Pseudapiconoma) preussi* (Aurivillius, 1904)
(Pl. II; ♂ Pl. 4; ♀ Pl. 23)

Short diagnosis. Imago. Forewing length: 18-30 mm. The only species in the subgenus with transparent markings on forewing. Female with transparent blotch between CuA$_1$-CuA$_2$ and with grey narrow band along distal margin of each tergite.

Male genitalia. Indistinguishable from those of *B. flavimacula* and *B. monotonia*.

Female genitalia. Similar to those of *B. flavimacula* and *B. monotonia* but signum reduced, weekly sclerotized, lateral lobes of distal margin of 8th sternite short.

Early stages. Unknown.

Biology. Unknown except for the collecting dates (January-April, August-December).

Distribution. Angola, Cameroon, Congo, DRC, Guinea, Nigeria, Tanzania.

Bergeria Kiriakoff, 1952

41. *Bergeria bourgognei* Kiriakoff, 1952
(Pl. II; ♂ Pl. 4)

Short diagnosis. Imago. Forewing length: 14-16 mm. This species and the similar *B. octava* form a pair that differ from all other *Bergeria* species in the colour of the hindwing which is yellow and ochraceous apically. Their forewing is ochraceous. The differences between these two species are small and restricted to a darker forewing and slightly wider dark apical band on the hindwing in *B. bourgognei*.

Male genitalia. Saccus slightly widened distally as in *B. octava*.

Female genitalia. Unknown.

Early stages. Unknown.

Biology. Unknown except for the collecting dates (December).

Distribution. DRC.

Remarks. The holotype is in fact a male, not a female as Kiriakoff (1952b) wrongly stated in his description. The neallotype male designated by him (Kiriakoff 1957a) is a male of *Lempkeella*. Only two males are known. *Bergeria fletcheri* Kiriakoff, 1957 is here proposed as a synonym of *B. bourgognei*. No differences were found to separate them.

42. *Bergeria haematochrysia* Kiriakoff, 1952
(Pl. II; ♂ Pl. 5; ♀ Pl. 24)

Short diagnosis. Imago. Forewing length: 16-23 mm. The only representative of the genus with the head, thorax, and abdomen entirely yellow, without any trace of red or brown.

Male genitalia. Saccus narrowed towards tip as in *B. ornata*, otherwise very similar to remaining representatives of the genus. The characters separating *B. haematochrysia* from *B. fletcheri* and *B. octava* presented by Kiriakoff (1957c, 1961a) are imperceptible.

51

Female genitalia. Ostium bursae two times as long as wide.

Early stages. Unknown.

Biology. Unknown except for the collecting dates (March, May, August, November, December).

Distribution. DRC, Cameroon, Rwanda, Zambia.

Remarks. In subspecies *B. h. occidentalis* the yellow, basal part of the female forewing is extended towards the middle part of the wing. This form is known only from one specimen.

43. *Bergeria octava* KIRIAKOFF, 1961
(Pl. II; ♂ Pl. 5)

Short diagnosis. Imago. Forewing length: 16 mm. Forewing pale brown, almost semi-transparent. For more diagnostic characters, see *B. bourgognei*.

Male genitalia. Saccus widened distally. The characters presented by KIRIAKOFF in the original description to separate *B. octava*, *B. haematochrysia*, and *B. fletcheri* (now syn. of *B. bourgognei*) are very small and subjective. They have to be verified on more extensive material.

Female genitalia. Unknown.

Early stages. Unknown.

Biology. Unknown except for the collecting dates (September).

Distribution. DRC.

Remarks. So far only the holotype male is known. This taxon is very similar to *B. bourgognei* and may be conspecific. Additional material is required to clarify its status.

44. *Bergeria ornata* KIRIAKOFF, 1959
(Pl. II; ♂ Pl. 5; ♀ Pl. 24)

Short diagnosis. Imago. Forewing length: 16-22 mm. Relatively large species with reddish forewing, yellow hindwing and head, and reddish yellow thorax and abdomen. Apical part of female hindwing black.

Male genitalia. Saccus slightly narrowed towards tip, otherwise resembling that of the other *Bergeria* species.

Female genitalia. Ostium bursae elongate, more than two times as long as wide.

Early stages. Unknown.

Biology. Unknown except for the collecting dates (April, June, July, September, October).

Distribution. DRC.

45. *Bergeria schoutedeni* KIRIAKOFF, 1952
(Pl. II; ♀ Pl. 24)

Short diagnosis. Imago. Forewing length: 19-21 mm. The species can be separated at once from all other *Bergeria* species by the mostly red thorax and abdominal tergites, and by the colouration of the hindwing which is yellow in basal 1/3 and then brownish black.

Male genitalia. Not examined.

Female genitalia. Sclerotized part of ductus bursae relatively short and wide.

Early stages. Not examined.

Biology. Unknown except for the collecting dates (January, February, April, May, August-December).

Distribution. DRC.

Remarks. Both sexes are similarly coloured. Easy to recognize without examining the genitalia because some external diagnostic features are located on the abdomen. The single available male specimen was not dissected.

46. *Bergeria tamsi* KIRIAKOFF, 1952
(Pl. II)

Short diagnosis. Imago. Forewing length: 17 mm. Wings unicolorous, brownish black; tergites brown with red bands, without yellow scales.

Male genitalia. Unknown.

Female genitalia. Not examined.

Early stages. Unknown.

Biology. Unknown except for the collecting dates (August).

Distribution. DRC.

Remarks. So far only the holotype female is known. The species is easily distinguished from the remaining *Bergeria* species without examining copulatory organs. Moreover, some external diagnostic features are located on the abdomen and therefore it was not dissected.

Cameroonia PRZYBYŁOWICZ [gen. nov.]

47. *Cameroonia nigriceps* (AURIVILLIUS, 1904)
(Pl. II; ♂ Pl. 5; ♀ Pl. 24)

Short diagnosis. Imago. Forewing length: 16-19 mm. Very characteristic species. Brightly, rust-red with dark brown veins and forewing markings, cilia concolorous with veins; male antenna with long setae.

Male genitalia. Phallus short, very wide; valva short, extremely wide; distal process of costa directed inwardly.

Female genitalia. Ductus bursae large, membranous, elongate, wide; corpus bursae small.

Early stages. Unknown.

Biology. Unknown except for the collecting dates (Mai, September-November).

Distribution. Cameroon.

Hippurarctia KIRIAKOFF, 1953

48. *Hippurarctia cinereoguttata* (STRAND, 1912)
(Pl. II; ♂ Pl. 5; ♀ Pl. 25)

Short diagnosis. Imago. Forewing length: 22-27 mm. Forewing ochraceous with pink blotches in middle; hindwing pink. Otherwise similar to *H. judith*.

Male genitalia. Indistinguishable from congeners.

Female genitalia. Bursa copulatrix with very dense, concentric plicae provided with small, chitinized thorns; ostium bursae situated into very deep concavity of distal margin of 8th sternum.

Early stages. Unknown.

Biology. Unknown except for the collecting dates (January).

Distribution. Congo, Equatorial Guinea.

Remarks. The description of the male genitalia presented by KIRIAKOFF (1960b: 28) does not emphasize the differences but rather show the similarities with previously described species.

The male of *H. cinereoguttata* was described for the first time by OBERTHÜR (1911: 469), but incorrectly determined as *Metarctia haematoessa* HOLLAND. This mistake was noted by HAMPSON (1914a: 65) who described these males as *Metarctia cameruna*. *Metarctia cameruna* was synonymized with *M. cinereoguttata* by KIRIAKOFF (1959a: 188).

49. *Hippurarctia ferrigera* (DRUCE, 1910)
(Pl. II; ♂ Pl. 5; ♀ Pl. 25)

Short diagnosis. Imago. Forewing length: 20-26 mm. Forewing dark brown with paler, irregular markings; distal part usually paler than proximal. Hindwing pale, brownish yellow. The species mostly resembles *H. taymansi*.

Male genitalia. Indistinguishable from those of the other representatives of the genus (see **Remarks** under *H. cinereoguttata*).

Female genitalia. Bursa copulatrix with delicate, concentric plicae; position of ostium bursae intermediate between that in *H. cinereoguttata* and *H. judith*.

Early stages. Unknown.

Biology. Unknown except for the collecting dates (January, March-May, July-October). In Kakamega Forest (Kenya) moths have been collected at an artificial light source located inside the middle-aged and young secondary forest.

Distribution. Cameroon, DRC, Kenya, Uganda.

50. *Hippurarctia judith* KIRIAKOFF, 1959
(Pl. II; ♂ Pl. 5; ♀ Pl. 25)

Short diagnosis. Imago. Forewing length: 19-27 mm. This is the only representative of the genus with pink hindwing and dark, brownish black forewing. *Hippurarctia cinereoguttata* can be confused with *H. judith* but differs in the prominent pale pink blotches in the middle of the forewing.

Male genitalia. Indistinguishable from those of the other *Hippurarctia* (see **Remarks** under *H. cinereoguttata*).

Female genitalia. Corpus bursae without plicae; ostium bursae in shallow concavity at distal margin of 8th sternum.

Early stages. Unknown.

Biology. Unknown except for the collecting dates (January, March, April, July).

Distribution. DRC.

51. *Hippurarctia taymansi* (ROTHSCHILD, 1910)
(Pl. II; ♂ Pl. 5)

Short diagnosis. Imago. Forewing length: 21-25 mm. Forewing pale brown with dark brown, contrasting markings. Hindwing pale, brownish yellow.

Male genitalia. Indistinguishable from congeners (see **Remarks** under *H. cinereoguttata*).

Female genitalia. Unknown.

Early stages. Unknown.

Biology. Unknown except for the collecting dates (May, July).

Distribution. DRC.

Lempkeella KIRIAKOFF, 1953

52. *Lempkeella avellana* (KIRIAKOFF, 1957)
(Pl. III; ♂ Pl. 5)

Short diagnosis. Imago. Forewing length: 15 mm. The holotype male is worn and faded but the forewing is more reddish than in the other two species.

Male genitalia. Apical process of costa of valva shorter and more flattened than in other *Lempkeella* species.

Female genitalia. Unknown.

Early stages. Unknown.

Biology. Unknown except for the collecting dates (April, September).

Distribution. DRC.

Remarks. The species is here transferred from *Bergeria*. Its relationships to *L. dufranei* and *L. vanoyei* remains uncertain as all species are similar to each other and the female genitalia are unknown.

53. *Lempkeella dufranei* (KIRIAKOFF, 1952)
(Pl. III; ♂ Pl. 6)

Short diagnosis. Imago. Forewing length: 13 mm. It can be separated from *L. vanoyei* by the yellow anal part of the hindwing and more apically pointed forewing. From *L. avellana* it differs in having a darker forewing.

Male genitalia. Apical process of costa of valva longer than in *avellana*; no differences from *L. vanoyei* were found.

Female genitalia. Unknown.

Early stages. Unknown.

Biology. Unknown except for the collecting dates (September).

Distribution. DRC (Elisabethville).

Remarks. Only four males (in poor condition) are known. Neallotype female was designated in the collection of RMCA by KIRIAKOFF (unpublished), but this specimen is a male of *Bergeria bourgognei* KIRIAKOFF, 1952b.

54. *Lempkeella vanoyei* (KIRIAKOFF, 1952)
(Pl. III)

Short diagnosis. Imago. Forewing length: 13 mm. Very dark species; anal part of hindwing with just a few pale scales.

55

Male genitalia. Indistinguishable from *L. dufranei*.

Female genitalia. Unknown.

Early stages. Unknown.

Biology. Unknown except for the collecting dates (July).

Distribution. DRC.

Remarks. More material is needed for a proper interpretation of the taxon. The drawing of the valva presented in the original description is incorrect due to the wrong interpretation of the original genital slide. The quality of the holotype male preparation made by KIRIAKOFF is not sufficient for reproducing it here.

Mecistorhabdia KIRIAKOFF, 1953

55. *Mecistorhabdia haematoessa* (HOLLAND, 1893)
(Pl. III; ♂ Pl. 6; ♀ Pl. 25)

Short diagnosis. Imago. Forewing length: 18-23 mm. Forewing rust brown with blackish markings. The colouration varies from almost entirely pale to very dark rust brown.

Male genitalia. Valva short; transtilla prominent, provided with few stout setae apically; distal process of costa much elongate, narrow, setose in terminal 1/3; sacculus in form of elongate, narrow process; phallus straight, vesica in basal part with one small thorn and two opposite, membranous pouches.

Female genitalia. Antevaginal and postvaginal plates forming pair of prominent, sclerotized concavities surrounding ostium bursae lateroventrally; ostium bursae in form of short, narrow, sclerotized vestibule.

Early stages. Unknown.

Biology. Unknown except for the collecting dates (January, March-May, July-October). In Kakamega Forest (Kenya) moths have been collected at an artificial light source located inside the old secondary forest.

Distribution. Cameroon, DRC, Gabon, Kenya, Rwanda, Uganda, Zambia.

Remarks. *Mecistorhabdia burgessi* KIRIAKOFF, 1957 is synonymized with *M. haematoessa* by PRZYBYŁOWICZ & KÜHNE 2008: 150. According to KIRIAKOFF (1957c) it is an eastern representative of *Mecistorhabdia*. However, new findings from Rwanda do not confirm the distinctness of these two taxa. The habitus and genital morphology of the holotype of *M. burgessi* match the intraspecific variability of *M. haematoessa*.

Melisa WALKER, 1854

56. *Melisa croceipes* (AURIVILLIUS, 1892)
(Pl. III; ♂ Pl. 6; ♀ Pl. 26)

Short diagnosis. Imago. Forewing length: 19-23 mm. Small species. Male with Cu_1 and Cu_2 of forewing parallel to each other; forewing underside with prominent streak of semi-erect, paler, androconial scales between M_1 and M_2; $A1$-$A2$ long; tuft of orange hair at apex of abdomen absent. Female with foretibia black basally.

Male genitalia. Uncus divided terminally into two lobes, longer than in *M. diptera*; distal process of costa short; saccus triangular; phallus very short and broad; spines in vesica short.

Female genitalia. Colliculum strongly sclerotized; ductus seminalis narrow from the base.

Early stages. Unknown.

Biology. Unknown. Adults collected throughout the year except for the collecting dates (May, June).

Distribution. Burundi, Cameroon, DRC, Ghana, Ivory Coast, Nigeria, Tanzania.

57. *Melisa diptera* (Walker, 1854)
(Pl. III; ♂ Pl. 6; ♀ Pl. 26)

Short diagnosis. Imago. Forewing length: 23-27 mm. Male with Cu_1 and Cu_2 of forewing very short, divergent; forewing underside without streak of semi-erect, paler, androconial scales between M_1 and M_2; A1-A2 very short; abdomen with tuft of orange hair apically. Foretibia of female orange, without any trace of black scales.

Male genitalia. Uncus divided terminally into two lobes, shorter than in *M. croceipes*; distal process of costa long; saccus weakly developed, evenly rounded; phallus elongate; spines in vesica longer than in *M. croceipes*.

Female genitalia. Colliculum not sclerotized; ductus seminalis broadened towards opening.

Early stages. Unknown.

Biology. Unknown. Adults collected throughout the year.

Distribution. Cameroon, DRC, Gabon, Ivory Coast, Nigeria.

58. *Melisa hancocki* Jordan, 1936
(Pl. III)

Short diagnosis. Imago. Forewing length: 23-25 mm. Very similar to *M. croceipes* but head, thorax, and legs entirely black with bluish shine.

Male genitalia. Not examined.

Female genitalia. Not examined.

Early stages. Unknown.

Biology. Unknown except for the collecting dates (October).

Distribution. Uganda.

Remarks. Known from a single pair. The species mostly resembles *M. croceipes* and it is possibly only a colour form. However, the type locality is the easternmost for the genus. The other two species are abundant in the Congo Basin, but do not extend far to the east. New material is necessary to clarify the position of this taxon. Because of the presence of important characters on the abdominal segments the genitalia were not dissected.

Melisoides Strand, 1912

59. *Melisoides lobata* Strand, 1912
(Pl. III; ♂ Pl. 6; ♀ Pl. 26)

Short diagnosis. Imago. Forewing length: 17-23 mm. *Melisoides lobata* is most similar to representatives of *Melisa* and *Paramelisa*. It differs from *Melisa* species by the presence of narrow, transverse yellow stripes on the abdominal tergites. From *Paramelisa* it may be separated easily by the glossy forewing colouration.

Male genitalia. An elongate saccus separates this species from *Melisa* and *Paramelisa*.

Female genitalia. Differ from those of *Melisa* and *Paramelisa* by the presence of a pair of parallel, deep depressions on the antevaginal plate.

Early stages. Unknown.

Biology. Unknown except for the collecting dates (January - March, May, June, September, October, December).

Distribution. Cameroon, DRC, Equatorial Guinea, Ivory Coast.

Metamicroptera Hulstaert, 1923

60. *Metamicroptera christophi* Przybyłowicz, 2005
(Pl. III; ♂ Pl. 6)

Short diagnosis. Imago. Forewing length: 14 mm. Patches on forewing much smaller than in *M. rotundata*; frons white; eye small, flattened; thorax and abdominal tergites with short, flat, adherent scales; proximal part of coxae white.

Male genitalia. Cornuti about ten times longer than wide, much longer than width of phallus.

Female genitalia. Unknown.

Early stages. Unknown.

Biology. Unknown except for the collecting dates (November, December).

Distribution. Zambia, Tanzania.

Remarks. Only two males are known so far.

61. *Metamicroptera rotundata* Hulstaert, 1923
(Pl. III; ♂ Pl. 6)

Short diagnosis. Imago. Forewing length: 16-18 mm. Frons black; eyes prominent, convex; thorax and abdominal tergites with hair-like scales; coxae of all legs black.

Male genitalia. Cornuti shorter than width of terminal part of phallus.

Female genitalia. Unknown.

Early stages. Unknown.

Biology. Unknown except for the collecting dates (January – May, September – December).

Distribution. DRC, Zambia.

Remarks. The species is common in the known range but only males are known until now. There is strong evidence that genera *Metamicroptera* and *Pseudmelisa* artificially separate males and females (respectively) of the same entities.

Metarctia Walker, 1855

Subgenus *Collocaliodes* Kiriakoff, 1957

62. *Metarctia (Collocaliodes) aethiops* (Kiriakoff, 1973)

Short diagnosis. Imago. Forewing length: 13-15 mm. According to Kiriakoff (1973a) this species is easy to separate from the remaining species of the subgenus (except *M. fuliginosa*) by the dark hindwing. From *M. fuliginosa* it differs in the slightly narrower forewing, shorter pectination of antenna, darker head, and buff orange (not grey-brown) abdomen (Kiriakoff 1973).

Male genitalia. According to Kiriakoff (1973) it can be separated by the wider uncus (in lateral view), angular (not rounded) costal margin of basal process of valva, and relatively much longer phallus.

Female genitalia. Unknown.

Early stages. Unknown.

Biology. Unknown except for the collecting data (May).

Distribution. Malawi (Mkuwadzi Forest, Nkata Bay).

Remarks. Described and known so far from two males collected at the same locality. The type housed in the BMZ (pers. comm. V. MUYAMBO) is inaccessible for study.

63. *Metarctia (Collocaliodes) collocalia* KIRIAKOFF, 1957
(Pl. III; ♂ Pl. 6)

Short diagnosis. Imago. Forewing length: 16-18 mm. Nominotypical subspecies pale fuscous with lighter basal half of hindwing; ssp. *kilimaensis* generally much darker; ssp. *montium* with hindwing pale pink.

Male genitalia. Basoventral process of valva prominent, well sclerotized, wrinkled; terminal part of valva relatively short and wide; saccus and basal part of phallus lacking. See **Remarks**.

Female genitalia. Unknown.

Early stages. Unknown.

Biology. Unknown except for the collecting dates (February, May, June, August, September, November). In Kakamega Forest (Kenya) *M. c. montium* has been collected at an artificial light source located both inside the young secondary forest and in open habitat (farmland).

Distribution. DRC, Kenya, Malawi, Tanzania, Zimbabwe.

Remarks. The taxonomic status of the subspecies and of some other similar taxa (*M. aethiops*, *M. fuliginosa*, *M. margaretha*, *M. pavlitzkae*) is not clear. The whole subgenus should be revised based on extensive material to enable the proper interpretation of the differences between specimens and populations.

Based on the present stage of knowledge of the subgenus, the morphology of the genitalia can not be used to separate species. Some characters are unique for the subgenus,

 e.g. the distinct, sclerotized process at the basoventral part of the valva; the straight and narrow phallus; the narrow and elongate vesica devoid of any sclerotized structures and provided with a distinct pouch-like process. The shape of the valva, uncus, and saccus vary significantly between specimens and the characters pointed out as significant in original descriptions can not be treated as such. Until a revision based on extensive material is made all generalizations are highly subjective. Therefore, the short information on genital morphology presented here refers to particular, illustrated specimens (mostly holotypes).

64. *Metarctia (Collocaliodes) debauchei* KIRIAKOFF, 1953
(Pl. III; ♂ Pl. 6)

Short diagnosis. Imago. Forewing length: 17 mm. The species differs from all other *Collocaliodes* species (except *M. jansei*) in the pale pink hindwing.

Male genitalia. Holotype indistinguishable from the other species except for exceptionally large, wide, triangular saccus. See **Remarks** for *M. collocalia*.

Female genitalia. Unknown.

Early stages. Unknown.

Biology. Unknown except for the collecting dates (April).

Distribution. Burundi.

Remarks. The structure of the male genitalia is typical of members of subgenus *Collocaliodes* to which the species is here moved. The position and the rank of this taxon within the group are unclear. *Metarctia debauchei* is very similar to *M. collocalia montium* and *M. jansei*. The genitalia do not show any characters that could distinguish it from its congeners. Despite all these similarities our present state of knowledge of this group of similar taxa does now allow for formal synonymizations.

65. *Metarctia (Collocaliodes) dracoena* KIRIAKOFF, 1953
(Pl. III; ♂ Pl. 7)

Short diagnosis. Imago. Forewing length: 13-14 mm. Forewing brown with small, darker, elongate blotch at end of cell; hindwing pale cream, anal region suffused with pinkish; thorax concolorous with forewing; head, patagia and upper part of abdomen rust brown. Most similar to *M. jansei* and *M. seydeliana* from which it can be separated on the basis of genitalic characters.

Male genitalia. Basoventral process of valva prominent, well sclerotized; terminal part of valva very short and wide; saccus narrow, elongate. See **Remarks** for *M. collocalia*.

Female genitalia. Unknown.

Early stages. Unknown.

Biology. Unknown except for the collecting dates (May).

Distribution. DRC.

Remarks. The type series consists of two rather worn specimens. The GS of holotype lack the phallus.

66. *Metarctia (Collocaliodes) fuliginosa* KIRIAKOFF, 1953
(Pl. III; ♂ Pl. 7)

Short diagnosis. Imago. Forewing length: 12 mm. The darkest species of the subgenus. Hindwing only slightly paler than forewing, darker than abdomen, paler at base; dark blotch on forewing absent. May be confused with *M. tenebrosa* or *M. pavlitzkae*.

Male genitalia. As *M. dracoena* but saccus lacking. See **Remarks** for *M. collocalia*.

Female genitalia. Unknown.

Early stages. Unknown.

Biology. Unknown except for the collecting dates (April).

Distribution. DRC (Sankuru: Lusambo).

Remarks. Described and known from one male only.

67. *Metarctia (Collocaliodes) jansei* KIRIAKOFF, 1957
(Pl. III; ♂ Pl. 7)

Short diagnosis. Imago. Forewing length: 16-17 mm. Externally indistinguishable from *M. dracoena*.

Male genitalia. Basoventral process of valva moderate, well sclerotized; terminal part of valva very short and wide; saccus wide basally, elongate, triangular; phallus slightly curved in terminal third. See **Remarks** for *M. collocalia*.

Female genitalia. Unknown.

Early stages. Unknown.

Biology. Unknown.

Distribution. RSA.

Remarks. This is the only representative of the subgenus inhabiting the southernmost areas of the African continent. Its taxonomic status is doubtful. See **Remarks** under *M. dracoena*.

68. *Metarctia (Collocaliodes) olbrechtsi* Kɪʀɪᴀᴋᴏꜰꜰ, 1953
(Pl. III; ♂ Pl. 7)

Short diagnosis. Imago. Forewing length: 12 mm. Smaller and darker than *M. seydeliana* and *M. dracoena*, very similar to *M. fuliginosa*. Dark blotch on forewing indistinct; hindwing paler than forewing, entirely suffused with brownish scales; head, patagia and abdomen concolorous with hindwing.

Male genitalia. Basoventral process of valva large, well sclerotized; terminal part of valva very short and wide; saccus intermediate between *M. dracoena* and *M. tenebrosa*; phallus straight. See **Remarks** for *M. collocalia*.

Female genitalia. Unknown.

Early stages. Unknown.

Biology. Unknown except for the collecting dates (March).

Distribution. DRC.

Remarks. Known from two males only.

69. *Metarctia (Collocaliodes) pavlitzkae* Kɪʀɪᴀᴋᴏꜰꜰ, 1961
(Pl. III; ♂ Pl. 7)

Short diagnosis. Imago. Forewing length: 18 mm. Almost identical to *M. fuliginosa* and *M. tenebrosa* but larger, with rust brown abdomen and paler base of hindwing without pink tinge.

Male genitalia. Very similar to those of *M. tenebrosa*. See **Remarks** for *M. collocalia*.

Female genitalia. Unknown.

Early stages. Unknown.

Biology. Unknown except for the collecting dates (November).

Distribution. Tanzania (Sakarani).

Remarks. Only the holotype male is known. It could represent a colour form of *M. fuliginosa* or *M. tenebrosa*.

70. *Metarctia (Collocaliodes) seydeliana* Kɪʀɪᴀᴋᴏꜰꜰ, 1953
(Pl. III; ♂ Pl. 7)

Short diagnosis. Imago. Forewing length: 18 mm. Resembling *M. dracoena* but more yellowish in general appearance.

Male genitalia. Very similar to *M. olbrechtsi* but basoventral process of valva larger and saccus very narrow, elongate. See **Remarks** for *M. collocalia*.

Female genitalia. Unknown.

Early stages. Unknown.

Biology. Unknown except for the collecting dates (February).

Distribution. DRC.

Remarks. Determination is possible on the basis of the male genitalia, but see **Remarks** under *M. dracoena*.

The male labeled in the collection as holotype and depicted in the original description as holotype bears the label with collecting date: 21.02.1934. In the text of the description the collecting dates of holotype and paratype are interchanged.

71. *Metarctia (Collocaliodes) tenebrosa* Le Cerf, 1922
(Pl. III; ♂ Pl. 7; ♀ Pl. 26)

Short diagnosis. Imago. Forewing length: 14-16 mm. Very dark, small species resembling *M. fuliginosa*, but larger.

Male genitalia. Basoventral process of valva large, well sclerotized; terminal part of valva relatively long and narrow; saccus prominent, wide basally, elongate, triangular; phallus narrow, elongate, straight. See **Remarks** for *M. collocalia*.

Female genitalia. Sterigma heavily sclerotized; ostium in form of narrow, transverse slit; colliculum weakly sclerotized (almost membranous), delicately plicate; signum absent.

Early stages. Unknown.

Biology. Unknown except for the collecting dates (January, April).

Distribution. Kenya.

Remarks. This is the only species in the subgenus for which the female is known.

The information presented both in original description and on labels attached to the holotype not enable the precise definition of the type locality.

Subgenus *Hebena* Walker, 1856

72. *Metarctia (Hebena) cinnamomea* (Wallengren, 1860)
(Pl. III; ♂ Pl. 7)

Short diagnosis. Imago. Forewing length: 26-27 mm. Similar in size to *M. rubra* but entirely pinkish-red except for black antennae, tarsi, and part of the tibia. *Metarctia subincarnata*, with which it might be confused, is much smaller.

Male genitalia. Transtilla relatively small; setae on valva delicate.

Female genitalia. Unknown.

Early stages. Unknown.

Biology. Unknown except for the collecting dates (November, December).

Distribution. RSA.

Remarks. Only the male is known.

73. *Metarctia (Hebena) henrardi* Kiriakoff, 1953
(Pl. III; ♂ Pl. 7; ♀ Pl. 27)

Short diagnosis. Imago. Forewing length: 13-17 mm. Looks like a small *M. lateritia*, but margins on forewing much less contrasting and forewing often unicolorous pale brownish pink.

Male genitalia. Transtilla extremely large, terminal process of valva narrow, elongate, slightly upcurved. Indistinguishable from these of *M. subincarnata*.

Female genitalia. Ostium bursae narrow, creviced, forming postmedially prominent, membranous cavity surrounded by pair of large, sclerotized cavities; colliculum short, broad, with weakly sclerotized plate; signum absent.

Early stages. Unknown.

Biology. Unknown except for the collecting dates (March).

Distribution. DRC.

Remarks. The taxonomic status of this taxon remains unclear.

74. *Metarctia (Hebena) lateritia* (HERRICH-SCHÄFFER, 1850-1858)
(Pl. III; ♂ Pl. 8; ♀ Pl. 27)

Short diagnosis. Imago. Forewing length: 17-26 mm. Very variable in size. General colouration pinkish red with greyish black markings between veins of forewing. These markings are much less contrast than in *M. rubra*. Thorax mostly red, rarely with some admixture of brownish black hairs.

Male genitalia. Transtilla prominent with long row of dense, elongate setae; deep concavity between transtilla and terminal process of valva, which is elongate and evenly rounded apically.

Female genitalia. Antevaginal plate forming sclerotized protrusion surrounding proximal margin of the ostium bursae cavity; colliculum elongate, sclerotized; signum present.

Early stages. Unknown.

Biology. The caterpillar feeds on *Senecio* sp. (Asteraceae) (SEVASTOPULO 1975). Adults collected throughout the year. In Kakamega Forest (Kenya) moths have been collected at an artificial light source located in open habitat (farmland).

Distribution. Angola, Burundi, Cameroon, DRC, Kenya, Rwanda, Tanzania, RSA, Uganda, Zambia.

Remarks. One of the most common representatives of the Thyretini, it has been recorded from most African countries south of the Sahara, in different habitats.

75. *Metarctia (Hebena) rubra* (WALKER, 1856)
(Pl. III; ♂ Pl. 8)

Short diagnosis. Imago. Forewing length: 30-32 mm. Huge species separated from *M. lateritia* by mostly greyish black thorax, interspaces between forewing veins, and costa of hindwing; hindwing deep pink, contrasting with forewing.

Male genitalia. Similar to those of *M. cinnamomea*, but some specimens with transtilla slightly larger, setae stout, and terminal part of valva more elongate.

Female genitalia. Unknown.

Early stages. Unknown.

Biology. Unknown except for the collecting dates (February).

Distribution. Angola.

Remarks. Only males are known so far.

In the original description of *A. rubra*, two specimens (males) are mentioned. In the BMNH collection only one specimen, from the DREGE's collection was located. It is a very worn and faded male. To prevent any doubts as to the identity of *A. rubra* this male labeled "Cape G. Hope; 42-77" [DREGE's collection] is here selected and designated as LECTOTYPE.

76. *Metarctia (Hebena) subincarnata* Kiriakoff, 1954
(Pl. III; ♂ Pl. 8)

Short diagnosis. Imago. Forewing length: 15-16 mm. Entirely pink but paler and smaller than *M. cinnamomea*.
Male genitalia. Indistinguishable from *M. henrardi*.
Female genitalia. Unknown.
Early stages. Unknown.
Biology. Unknown except for the collecting dates (October).
Distribution. DRC.
Remarks. Only two males are known until now.

Subgenus *Metarctia* Walker, 1855

77. *Metarctia (Metarctia) alticola* Aurivillius, 1925
(Pl. III; ♂ Pl. 8)

Short diagnosis. Imago. Forewing length: 12-15 mm. Body rust brown; forewing suffused with brownish black scales, with two indistinct black blotches at end of cell and near inner margin. Hindwing slightly paler.
Male genitalia. Resembling *M. flora*, but saccus slightly shorter and narrower.
Female genitalia. Unknown.
Early stages. Unknown.
Biology. Unknown except for the collecting dates (October, December).
Distribution. Rwanda.
Remarks. After its description the taxon *Metarctia rufescens* var. *alticola* Aurivillius, 1925 was neither mentioned by Kiriakoff in his catalogue (Kiriakoff 1960) nor in any other publication. The morphology of the genitalia – as in many species of Thyretini – can not be used to separate the species with certainty. However, the colouration and pattern are diagnostic. The validity of small differences in the shape of saccus betwen *M. alticola* and *M. flora* should be checked on more extensive material.

78. *Metarctia (Metarctia) atrivenata* Kiriakoff, 1956
(Pl. III; ♂ Pl. 8)

Short diagnosis. Imago. Forewing length: 19-23 mm. Similar to *M. longipalpis* in size and colouration, but veins of forewing black.
Male genitalia. Distal process of costa elongate but shorter than valva, slightly widened, and upcurved terminally; ventral protrusion at ½ length of costa densely covered with setae; proximal part of costa convex, forming low but distinct protuberance; phallus elongate, narrow, almost straight.
Female genitalia. Unknown.
Early stages. Unknown.
Biology. Unknown except for the collecting dates (March).
Distribution. Tanzania.

79. *Metarctia (Metarctia) benitensis* HOLLAND, 1893
(Pl. III; ♂ Pl. 8; ♀ Pl. 27)

Short diagnosis. Imago. Forewing length: 12-18 mm. Uniformly coloured brownish black except for paler hindwing.

Male genitalia. Indistinguishable from those of *M. paremphares*, but see **Remarks** under that name (111).

Female genitalia. Resembling those of *M. paremphares*, but ductus bursae shorter and wider; weakly sclerotized colliculum almost half as long as ductus bursae; signum even more weakly sclerotized than in *M. paremphares*.

Early stages. The pupation takes place in a loose cocoon on the ground (STRAND 1912).

Biology. According to STRAND (1912) the caterpillar is polyphagous on different plants and decaying leaves. Adults collected in March, July and August.

Distribution. DRC, Equatorial Guinea, Uganda.

Remarks. There are 2 specimens (1♂, 1♀) labeled "Type" in the CMNH.

80. *Metarctia (Metarctia) brunneipennis* HERING, 1932
(Pl. III; ♀ Pl. 27)

Short diagnosis. Imago. Forewing length: 17 mm. Medium-sized species. Head, thorax, and abdomen pale rusty; antennae and legs brownish black; forewing pale brown with indistinct darker blotch at end of cell; hindwing paler, creame with concolorous cilia.

Male genitalia. Unknown.

Female genitalia. Ostium bursae creviced, sclerotized, situated in deep, membranous cavity; inner distal angles of 8th sternite projected ventrally; ductus bursae narrow, elongate; ductus seminalis from 2/3 of its length; signum absent.

Early stages. Unknown.

Biology. Unknown except for the collecting dates (October).

Distribution. DRC (Elisabethville).

Remarks. Described and known from one female.

81. *Metarctia (Metarctia) burra* (SCHAUS in SCHAUS & CLEMENTS, 1893)
(Pl. IV; ♂ Pl. 8; ♀ Pl. 28)

Short diagnosis. Imago. Forewing length: 11-19 mm. Resembles *M. fulvia* but forewing darker; with pale rusty pink blotch at end of cell; hindwing distinctly paler, pinkish creame.

Male genitalia. Valva very short, broad; distal process of valva short, terminated ventrally in minute, sclerotized tooth; ventral protrusion indistinct, setose, located at base of distal process of valva; sacculus prominent; phallus short, broad.

Female genitalia. Ductus bursae short; colliculum membranous, with subsquare, convex, sclerite proximal to opening of ductus seminalis; corpus bursae large, signum elongate in form of two parallel, narrow plates.

Early stages. Unknown.

Biology. Unknown except for the collecting dates (March, June, July, August).

Distribution. DRC, Guinea, Nigeria, Sierra Leone, Sudan.

Remarks. Three taxa are for the first time here synonymized with *M. burra*: *M. hector* KIRIAKOFF, 1959, *M. chryseis*

KIRIAKOFF, 1973, and *M. sudanica* KIRIAKOFF, 1973. The types of the first two taxa show no differences from the type of *M. burra*. The type specimens of *M. sudanica* are slightly different and may represent an eastern subspecies of *M. burra*.

The problem of the number of specimens in the type series of *M. burra* needs a short comment. The species was described from an unstated number of specimens. However, some other information suggest that there was only one. This one individual was found in the collection of AMNH. The specimen matches the sex, colouration, general appearance, and expanse of forewing as given in the description and on the colour plate. Taking all this into consideration I treat this specimen as a holotype and the only specimen of the type series of *M. burra*.

82. *Metarctia (Metarctia) burungae* DEBAUCHE, 1942
(Pl. IV; ♂ Pl. 8)

Short diagnosis. Imago. Forewing length: 16-17 mm. Forewing brown with darker basal and apical areas; middle part with prominent darker blotch near end of cell; hindwing pale brown.

Male genitalia. Distal process of costa elongate, slightly upcurved terminally; saccus wide, broadly rounded terminally.

Female genitalia. Unknown.

Early stages. Unknown.

Biology. Unknown except for the collecting dates (March, July, October).

Distribution. DRC.

83. *Metarctia (Metarctia) carmel* KIRIAKOFF, 1957
(Pl. IV; ♂ Pl. 8)

Short diagnosis. Imago. Forewing length: 12 mm. Very similar in colouration to *M. burra* and *M. fulvia* but can be separated from them by the absence of a dark blotch at the end of the cell.

Male genitalia. Very similar to these of *M. atrivenata* but ventral protrusion of costa less prominent; phallus much shorter, broad, straight. Indistinguishable from these of *M. burra*.

Female genitalia. Unknown.

Early stages. Unknown.

Biology. Unknown except for the collecting dates (April).

Distribution. Ethiopia.

Remarks. Described from a single male. Taxonomic status unclear.

84. *Metarctia (Metarctia) diversa* BETHUNE-BAKER, 1911
(Pl. IV; ♂ Pl. 9)

Short diagnosis. Imago. Forewing length: 19-22 mm. Forewing unicorous, pale brown, with concolorous cilia and indistinct darker dot at end of cell; costa yellow. Hindwing semi-transparent, pale pink.

Male genitalia. Distal process of costa narrow, elongate, slightly upcurved terminally; ventral protrusion elongate, covered with short setae, tipped with small appendix near margin of costa; saccus narrow; phallus long, narrow, somewhat S-shaped.

Female genitalia. Unknown.

Early stages. Unknown.

Biology. Unknown except for the collecting dates (February, March, October-December).

Distribution. Angola, DRC.

85. *Metarctia (Metarctia) fario* KIRIAKOFF, 1957
(Pl. IV; ♂ Pl. 9)

Short diagnosis. Imago. Forewing length: 11-14 mm. Similar to *M. negusi, M. pallens*, and *M. salmonea.* Characterized by dark, brownish black head, thorax, costa, and CuA vein until end of cell.

Male genitalia. Distal process of valva narrow, elongate, widened terminally, shorter than valva; phallus relatively short and broad.

Female genitalia. Unknown.

Early stages. Unknown.

Biology. Unknown except for the collecting dates (October).

Distribution. DRC.

86. *Metarctia (Metarctia) flaviciliata* HAMPSON, 1907
(Pl. IV; ♂ Pl. 9; ♀ Pl. 28)

Short diagnosis. Imago. Forewing length: 17-25 mm. One of the bigger species of the genus. Forewing uniformly brownish black with cream cilia; hindwing slightly paler with cream base and cilia; head and abdomen rust; antennae black; thorax brownish black. Female similar but larger and with hindwing darker.

Male genitalia. Distal process of valva narrow, elongate, upcurved terminally; ventral protrusion of costa medially situated, slightly elongate, with row of short setae; saccus short.

Female genitalia. Ventral and dorsal pheromone glands long, slender, with several processes terminally; postvaginal plate forming upraised sclerite; ductus bursae membranous; signum elongate, not divided into two parallel parts.

Early stages. A short description of the caterpillar and pupa has been provided by FONTAINE (1992).

Biology. In captivity the caterpillar feeds on *Pelargonium*, and the pupation lasts 22-24 days (FONTAINE 1992). According to LE PELLEY (1959) the host plants in Uganda are *Cineraria* sp. (Asteraceae), *Commelina* sp. (Commelinaceae) and *Pennisetum purpureum* (Poaceae). Adults collected in April, July, September and November.

Distribution. DRC, Uganda.

Remarks. The type series consist of two syntypes (♂, ♀). Both of them have been found in the BMNH. The male was selected and here designated as the LECTOTYPE to prevent any doubts as to the identity of this species. The female was labeled as PARALECTOTYPE.

The record from Uganda is cited from biological data published by LE PELLEY (1959). No specimens were found in collections.

87. *Metarctia (Metarctia) flavicincta* AURIVILLIUS, 1900
(Pl. IV; ♂ Pl. 9)

Short diagnosis. Imago. Forewing length: 11-15 mm. Small species separated from *M. sarcosoma* and *M. fontainei* by combination of blackish brown thorax and abdominal sterna.

Male genitalia. Distal process of valva of moderate length, narrow, slightly upcurved; dorsal protrusion and ventral process of valva absent; saccus elongate; phallus narrow; vesica with finger-like membranous pouch.

Female genitalia. Unknown.

Early stages. Unknown.

Biology. The caterpillar was observed in Kenya feeding on "pasture grasses" (LE PELLEY 1959). Adults collected in February-April, June-August and November-December. In Kakamega Forest (Kenya) moths have been collected at an artificial light source located mainly in open habitat (farmland) but also rarely inside the forest.

Distribution. Angola, DRC, Kenya, Uganda.

Remarks. The type locality "Mukimbungu" provided here was not part of the original description, but it is cited in the title of the paper of AURIVILLIUS (1900). The only information written on the labels is "Congo".

88. *Metarctia (Metarctia) flavivena* HAMPSON, 1901
(Pl. IV; ♂ Pl. 9; ♀ Pl. 28)

Short diagnosis. Imago. Forewing length: 15-21 mm. Head and thorax blackish brown, abdomen dark red; forewing blackish brown, veins and cilia cream; hindwing cream, costa blackish brown. Very variable species. Cream markings often much darker, up to pale red. The latter form looks like small *M. lateritia*, but can be separated at once by the always blackish brown thorax. The rare form *zegina* has both wings concolorous and pink veins and cilia.

Male genitalia. Valva short, wide; distal process short; transtilla forming prominent process directed ventrally; saccus elongate, tapering towards tip.

Female genitalia. Antevaginal plate forming projected, sclerotized, plate; colliculum moderately sclerotized; signum distinctly elongate, divided into two parallel parts.

Early stages. Unknown.

Biology. The caterpillar feeds on *Bidens pilosa*, *Dahlia* sp., and *Sonchus* sp. (Asteraceae) (SEVASTOPULO 1975). Adults collected in February-April, August-December. In Kakamega Forest (Kenya) moths have been collected at an artificial light source located inside the young secondary forest.

Distribution. Angola, DRC, Ethiopia, Kenya, Nigeria, Zimbabwe.

Remarks. A very common species. Several individual colour variations have been described.

89. *Metarctia (Metarctia) flora* KIRIAKOFF, 1957
(Pl. IV; ♂ Pl. 9)

Short diagnosis. Imago. Forewing length: 15 mm. Forewing fuscous with indistinct rusty blotch beyond cell and between veins CuA_2 and A1-A2. Hindwing paler.

Male genitalia. Valva short, terminal process of costa long, as long as valva; saccus unusually large, as long as wide, evenly rounded.

Female genitalia. Unknown.

Early stages. Unknown.

Biology. Unknown except for the collecting dates (December).

Distribution. Rwanda.

90. *Metarctia (Metarctia) fontainei* KIRIAKOFF, 1953
(Pl. IV; ♂ Pl. 9; ♀ Pl. 28)

Short diagnosis. Imago. Forewing length: 20 mm. Very similar to *M. sarcosoma* from which it differs by the dark brown colour of the thorax.

Male genitalia. Distal process of valva very short, broad, outer margin covered with numerous, long setae; ventral protrusion densely covered with long setae and adjacent to dorsal margin of costa; saccus triangular; phallus short, slightly broadened in terminal part.

Female genitalia. Postvaginal plate in form of single, rounded, shallow cavity densely covered with short spines; ductus bursae very short; colliculum with sclerotized plate.

Early stages. Unknown.

Biology. Unknown except for the collecting dates (March, December).

Distribution. DRC.

Remarks. The taxonomic status of this taxon requires further investigation. Possibly it is only a subspecies of *M. sarcosoma*.

91. *Metarctia (Metarctia) forsteri* KIRIAKOFF, 1955
(Pl. IV; ♂ Pl. 9)

Short diagnosis. Imago. Forewing length: 18 mm. Similar to *M. fontainei* and *M. sarcosoma* from which it can be separated by the dark cilia of the forewing; hindwing suffused with black scales along outer margin; forewing underside pale ochraceous cream in basal 1/3, remaining part dark ochraceous.

Male genitalia. Resembling *M. hulstaertiana* but proximal part of costa even more convex; saccus apically pointed; phallus narrow, elongate, slightly S-shape.

Female genitalia. Unknown.

Early stages. Unknown.

Biology. Unknown.

Distribution. Cameroon.

Remarks. Only one male known. It may be only an individual colour form of *M. sarcosoma*.

92. *Metarctia (Metarctia) fulvia* HAMPSON, 1901
(Pl. IV; ♂ Pl. 9; ♀ Pl. 29)

Short diagnosis. Imago. Forewing length: 15-21 mm. Very similar to *M. burra* but entire body uniformly pale yellow and wings semi-transparent. Small blotch at end of forewing cell indistinct or absent.

Male genitalia. Distal process of valva relatively short; ventral protrusion very long, from medio-basal portion of valva until dorsal part of costa at 2/3 of length; saccus very long and narrow.

Female genitalia. Ostium bursae creviced; distal portion of postvaginal plate concave, membranous; ductus bursae long, slender; signum absent.

Early stages. Unknown.

Biology. Unknown except for the collecting dates (March, April, November).

Distribution. Kenya.

Remarks. *M. fulvia* was described from three males. The specimen bearing a small, red-bordered, round label with

69

"Type" is selected and here designated as LECTOTYPE to avoid any doubts as to the identity of this species. The remaining two males were labeled as PARALECTOTYPES.

M. neaera is here synonymized with *M. fulvia*. This taxon was originally described from an unstated number of males. However, the specimen illustrated by Fawcett (1915) on Pl. I fig. 6 is in fact a female. The wingspan and colouration of described specimen also suggest it was a female. It appears the mark "♂" was used in the description by mistake. In the BMNH there is a pair of *M. neaera* that can be recognized as the SYNTYPES and on which Kiriakoff noted on labels that the characters mentioned by Fawcett (1915) refer to both male and female. To avoid any doubts as to the identity of this species the female was selected as LECTOTYPE of *M. neaera* and the male was labeled as PARALECTOTYPE.

93. *Metarctia (Metarctia) fusca* Hampson, 1901
(Pl. IV; ♂ Pl. 10; ♀ Pl. 29)

Short diagnosis. Imago. Forewing length: 16-21 mm. Very characteristic, dark species. Forewing unicolorously brownish black with three small, cream dots: two just beyond apex of cell and one between cell and inner margin.

Male genitalia. Valva short, broad; distal process of valva long, narrow; ventral protrusion small, situated in dorso-terminal portion of valva; saccus broad, rounded terminally; phallus long, narrow, straight.

Female genitalia. Antevaginal and postvaginal plates fused in form of small, sclerotized protuberance; colliculum "c"-shaped, outer wall weakly sclerotized; signum absent.

Early stages. Unknown.

Biology. Unknown except for the collecting dates (August, December).

Distribution. Kenya.

94. *Metarctia (Metarctia) galla* Rougeot, 1977
(Pl. IV; ♂ Pl. 10)

Short diagnosis. Imago. Forewing length: 13-15 mm. Uniformly brownish black with pale cream antennae. Differs from other similarly coloured species by the forewing shape which is relatively short and wide distally, resembling species of subgenus *Thyretarctia*.

Male genitalia. Very similar to those of *M. kumasina*. Valva short, broad; distal process of valva short; proximal part of valva weakly convex; saccus prominent with rounded tip; phallus relatively longer and narrower than in *M. kumasina*.

Female genitalia. Unknown.

Early stages. Unknown.

Biology. Unknown except for the collecting dates (November).

Distribution. Ethiopia.

Remarks. Two different type localities: "réserve de Dinsho" mentioned in the original description and " réserve de Balé" printed on the label attached to the holotype refere to the same area. Réserve de Balé is now the Bale Mountains National Park and Dinsho is a small village situated in the northern part of the park.

95. *Metarctia (Metarctia) haematricha* Hampson, 1905
(Pl. IV; ♂ Pl. 10)

Short diagnosis. Imago. Forewing length: 16-22 mm. Very similar to *M. rufescens*. Differs in absence of narrow, yellow stripe along subcostal vein.

Male genitalia. Costa of valva distinctly convex; distal process weakly developed; ventral process slightly elongate, setose; saccus triangular.

Female genitalia. Unknown.

Early stages. Unknown.

Biology. Unknown except for the collecting dates (June, September).

Distribution. Ethiopia.

96. *Metarctia (Metarctia) hebenoides* (KIRIAKOFF, 1973)

Short diagnosis. Imago. Forewing length: 19 mm. According to the original description "resembling *A. robusta*, with similar pattern but much vivid and markedly smaller".

Male genitalia. Not examined.

Female genitalia. Unknown.

Early stages. Unknown.

Biology. Unknown except for the collecting date (January).

Distribution. Malawi.

Remarks. Type material not examined. The type housed in the BMZ (pers. comm V. MUYAMBO) is inaccessible for study.

97. *Metarctia (Metarctia) hulstaertiana* KIRIAKOFF, 1953
(Pl. IV; ♂ Pl. 10)

Short diagnosis. Imago. Forewing length: 12-14 mm. Forewing brownish black, hindwing cream. Belongs to a group of several dark ochraceous, similarly coloured species.

Male genitalia. Distal process of costa elongate but shorter than valva; ventral protrusion at ½ length of costa; proximal part of costa convex, forming low but distinct protuberance; saccus gradually narrowed towards blunt tip; phallus short, wide.

Female genitalia. Unknown.

Early stages. Unknown.

Biology. Unknown.

Distribution. DRC.

Remarks. Examination of male genitalia is necessary for proper determination.

98. *Metarctia (Metarctia) inconspicua* HOLLAND, 1892
(Pl. IV; ♂ Pl. 10)

Short diagnosis. Imago. Forewing length: 13-16 mm. Head, thorax (partly), and abdomen yellow; mesothorax and forewing brownish black; hindwing cream-grey.

Male genitalia. Valva triangular; terminal process of costa wide, of moderate length; inner protrusion of costa elongate, weakly developed with row of long setae; phallus narrow, straight, relatively short; vesica narrow, elongate, with basal part membranous, slightly widened, than covered with numerous, minute, spinules, with distal part membranous.

71

Female genitalia. Unknown.

Early stages. Unknown.

Biology. In Kenya *Zea mays* L. (Poaceae) was recorded as a host plant (LE PELLEY 1959). According to SEVASTOPULO (1975) the caterpillar feeds also on *Aster* sp., *Zinna* sp. (Asteraceae), and *Ipomea* sp. and *Stictocardia* sp. (Convolvulaceae). Adults collected in April, June and September. In Kakamega Forest (Kenya) moths have been collected at an artificial light source located both inside the young secondary forest and in open habitats (farmland, grassland).

Distribution. Kenya, Tanzania.

99. *Metarctia (Metarctia) johanna* (KIRIAKOFF, 1979)
(Pl. IV; ♂ Pl. 10)

Short diagnosis. Imago. Forewing length: 12-13 mm. Similar in size and colouration to *M. flavicincta* but much darker; paler costal line not marked.

Male genitalia. Distal process of valva narrow, elongate; ventral protrusion marked as elongate, setose area; saccus prominent, wide, broadly tipped; phallus, slightly S-shaped.

Female genitalia. Unknown.

Early stages. Unknown.

Biology. Unknown except for the collecting dates (May).

Distribution. Nigeria.

Remarks. Possibly a darker, colour form of *M. flavicincta*.

100. *Metarctia (Metarctia) kumasina* STRAND, 1920
(Pl. IV; ♂ Pl. 10)

Short diagnosis. Imago. Forewing length: 14 mm. Forewing pale fuscous, semi-transparent, costa lightly pinkish; antennae, head, patagia, and abdomen rust yellow; thorax fuscous.

Male genitalia. Valva short, broad; distal process of valva short, triangular; proximal part of valva distinctly convex; saccus prominent, triangular; phallus short, broad.

Female genitalia. Unknown.

Early stages. Unknown.

Biology. Unknown except for the collecting dates (May).

Distribution. Ethiopia, Ghana,

Remarks. Difficult to distinguish from several similarly coloured species. Formal description of STRAND is based on the six males mentioned by HAMPSON (1914a: 72) as "*M. pallida* Subsp. 1.". All of them have been found in BMNH and one bearing small, rounded, red-bordered "TYPE" label has been selected and here designated as LECTOTYPE. This is to prevent any doubts as to the taxonomic distinctness of *M. kumasina* from *A. pallida*. Remaining males were labeled as PARALECTOTYPES.

101. *Metarctia (Metarctia) lindemannae* KIRIAKOFF, 1961
(Pl. IV; ♂ Pl. 10)

Short diagnosis. Imago. Forewing length: 13-14 mm. Similar to *M. rufescens* but smaller, with costa of forewing brownish red.

Male genitalia. Distal process of valva of moderate length; ventral protrusion elongate, narrow, covered with setose margin from base of valva until dorso median area of costa; saccus prominent, triangular.

Female genitalia. Unknown.

Early stages. Unknown.

Biology. Unknown except for the collecting dates (November).

Distribution. Tanzania.

Remarks. Except for the type series there are four more males in the ZSM. Taxonomic position unclear.

102. *Metarctia (Metarctia) longipalpis* HULSTAERT, 1923
(Pl. IV; ♂ Pl. 10)

Short diagnosis. Imago. Forewing length: 19-22 mm. Very distinctive, rather large *Metarctia* with long, black palpi. Entire body unicolorous, rust brown, hindwing slightly paler; forewing with distinctive, black, narrow blotch at end of cell.

Male genitalia. Valvae and uncus asymmetrical.

Female genitalia. Unknown.

Early stages. Unknown.

Biology. Unknown except for the collecting dates (November).

Distribution. DRC.

Remarks. The species is known only from one male. The strange (unknown in any other Thyretini) asymmetry of the genitalia is probably atypical for the species and may be caused by genetic factors or parasites.

103. *Metarctia (Metarctia) lugubris* GAEDE, 1926
(Pl. IV; ♂ Pl. 11)

Short diagnosis. Imago. Forewing length: 12-15 mm. Similar to a small *M. burungae*.

Male genitalia. Distal process of valva elongate, narrow, as long as valva; ventral protrusion located at ½ length of costa, weakly marked, covered with several setae; saccus short, broad, rounded terminally; phallus elongate, straight.

Female genitalia. Not examined.

Early stages. Unknown.

Biology. Unknown except for the collecting dates (February, May, July).

Distribution. DRC, Tanzania, Uganda.

Remarks. Too few specimens of *M. lugubris* and *M. burungae* are available to clarify the problem of their taxonomic position.

104. *Metarctia (Metarctia) maria* KIRIAKOFF, 1957
(Pl. IV; ♂ Pl. 11)

Short diagnosis. Imago. Forewing length: 12 mm. Head, thorax, and forewing fuscous, darker in basal, costal, and inner parts. Antennae and abdomen pale ochreous yellow; hindwing pinkish cream. Underside cream except for dark fuscous costa on both pairs of wings.

Male genitalia. Distal process of costa elongate, slightly upcurved terminally; ventral protrusion at ½ length of costa, densely setose; proximal part of costa somewhat convex; phallus short, broad, almost straight.

Female genitalia. Unknown.

Early stages. Unknown.

Biology. Unknown except for the collecting dates (May).

Distribution. Guinea.

Remarks. Separation from other species with fuscous forewing and creamy hindwing possible only after examining genitalia.

105. *Metarctia (Metarctia) metaleuca* HAMPSON, 1914
(Pl. IV; ♂ Pl. 11)

Short diagnosis. Imago. Forewing length: 11-15 mm. Whole body uniformly brownish black except creamy hindwing. Colouration does not separate it from other similarly coloured species, e.g. *M. benitensis*, *M. hulstaertiana* or *M. paremphares*.

Male genitalia. Distal process of costa shorter than valva; broad basally; proximal part of costa somewhat convex; saccus of moderate size, rounded terminally; phallus short, broad.

Female genitalia. Unknown.

Early stages. Unknown.

Biology. Unknown except for the collecting dates (December).

Distribution. Cameroon, Liberia.

Remarks. The taxonomic rank and validity of several similarly looking, dark coloured species remain uncertain. Proper determination is possible only after examination of genitalia.

106. *Metarctia (Metarctia) morag* KIRIAKOFF, 1957
(Pl. IV; ♂ Pl. 11)

Short diagnosis. Imago. Forewing length: 11-14 mm. Small species, resembling *M. burungae* and *M. lugubris* from which it can be separated at once by the uniformly pale cream pinkish brown hindwing with blackish brown cilia.

Male genitalia. Distal process of costa rather short; dorso-ventral protrusion weak, setose; saccus wide, broadly tipped.

Female genitalia. Unknown.

Early stages. Unknown.

Biology. Unknown except for the collecting dates (February, June).

Distribution. DRC.

107. *Metarctia (Metarctia) negusi* KIRIAKOFF, 1957
(Pl. IV; ♂ Pl. 11)

Short diagnosis. Imago. Forewing length: 15 – 18 mm. Similar to *M. pallens* and *M. salmonea*. Forewing pale yellow-brown; costal margin, end of cell and space between inner margin and cell suffused with brown scales. Hindwing distinctly paler, creamy. Thorax brown, darker than head and abdomen.

Male genitalia. Distal process of costa shorter than valva; ventral protruding located in ½ length of costa; proximal part of costa convex.

Female genitalia. Unknown.

Early stages. Unknown.

Biology. Unknown.

Distribution. Ethiopia.

Remarks. So far known from 4 males only.

108. *Metarctia (Metarctia) nigritarsis* BERIO, 1943
(Pl. IV; ♂ Pl. 11; ♀ Pl. 29)

Short diagnosis. Imago. Forewing length: 15-19 mm. Forewing brownish black, hindwing paler. Belongs to the group of several dark, externally indistinguishable species.

Male genitalia. Distal process short, triangular, upcurved terminally; proximal portion of costa convex; ventral protrusion forming very distinct sharp process located at 2/3 length of costa; saccus elongate, triangular; phallus of moderate length, straight.

Female genitalia. Distal margin of 7th sternite concave; midanterior portion of antevaginal plate atrophied; ductus bursae membranous, straight, not very long; signum absent.

Early stages. Unknown.

Biology. Unknown except for the collecting dates (October).

Distribution. Eritrea.

109. *Metarctia (Metarctia) noctis* DRUCE, 1910
(Pl. IV; ♂ Pl. 11)

Short diagnosis. Imago. Forewing length: 13-14 mm. Very distinctive species slightly resembling *A. bicolora*. Head, antennae, patagia, tegulae, and underside of abdomen yellow; thorax and abdomen brown dorsally; wings blackish brown with concolorous cilia except for yellow costa and base of hindwing.

Male genitalia. Distal process of costa short, broad; proximal part of costa distinctly convex; saccus elongate, narrow, longer than distal process of costa.

Female genitalia. Unknown.

Early stages. Unknown.

Biology. Unknown.

Distribution. Ethiopia.

Remarks. So far only known from one male.

110. *Metarctia (Metarctia) pallens* BETHUNE-BAKER, 1911
(Pl. IV)

Short diagnosis. Imago. Forewing length: 13-16 mm. Very similar to *M. fario* except for rust-yellow head, thorax, and abdomen. Forewing dark cream suffused with dark ochraceous scales, dark markings paler.

Male genitalia. Not examined.

Female genitalia. Unknown.

Early stages. Unknown.

Biology. Unknown except for the collecting dates (November).

Distribution. Angola.

Remarks. The unique male (holotype) was not dissected. The differences in colouration and size between *M. pallens* and *M. fario* allow for an easy separation of these two species. Records from DRC (KIRIAKOFF 1953b) refere to the wrongly determined specimens.

111. *Metarctia (Metarctia) paremphares* HOLLAND, 1893
(Pl. IV; ♂ Pl. 11; ♀ Pl. 29)

Short diagnosis. Imago. Forewing length: 12-17 mm. One of the smaller species. Forewing uniformly brownish black; hindwing cream white. Thorax concolorous with forewing; head and abdomen brownish yellow.

Male genitalia. Terminal process of costa elongate, gradually narrowed towards apex; minute protrusion at 1/3 length of costa; inner process of costa absent; saccus short, wide, rounded apically; phallus narrow, straight, relatively short; vesica widened basally.

Female genitalia. Ostium bursae surrounded by numerous, long setae; colliculum weakly sclerotized; ductus bursae slightly shorter than bursa copulatrix; signum narrow, elongate, weakly sclerotized.

Early stages. Unknown.

Biology. Unknown except for the collecting dates (January, February, April, July, September and October). In Kakamega Forest (Kenya) moths have been collected at an artificial light source located mainly inside the young secondary forest and in open habitat (farmland).

Distribution. Angola, Gabon, Kenya.

Remarks. The species was described from the unstated number of specimens. As provided in the last line of the description, the "expanse" of the male and female may suggest that there were only two specimens (SYNTYPES). However in the CMNH three specimens labeled "Type" have been found. The additional male is smaller but otherwise shows no differences from the SYNTYPE male. The genitalia of both of them differ only insignificantly. The genitalia of *M. benitensis* are intermediate between those of both males of *M. paremphares*. However, the superficial differences and the lack of comparative material do not allow for the synonymization of these two taxa.

112. *Metarctia (Metarctia) paulis* KIRIAKOFF, 1961
(Pl. IV; ♂ Pl. 11)

Short diagnosis. Imago. Forewing length: 16 mm. Forewing brownish black, hindwing paler. Belongs to the group of several dark, externally indistinguishable species.

Male genitalia. Valva not shortened, distal process small, triangular, slightly upcurved at tip; ventral process well marked, covered with long setae; saccus small, triangular.

Female genitalia. Unknown.

Early stages. Unknown.

Biology. Unknown except for the collecting dates (July).

Distribution. DRC.

113. *Metarctia (Metarctia) phaeoptera* Hampson, 1909
(Pl. IV; ♂ Pl. 12; ♀ Pl. 30)

Short diagnosis. Imago. Forewing length: 10-14 mm. One of the smallest species. With relatively narrow, semi-transparent wings, looking very fragile. Wings brown; cilia slightly paler; apical area of forewing not semi-transparent. Thorax and antennae concolorous with wings; head, patagia, and abdomen yellow.

Male genitalia. As *M. atrivenata*, but distal process of costa longer.

Female genitalia. Ostium bursae large, ovate, located in membranous cavity; ductus bursae long, slender, membranous; signum absent.

Early stages. Unknown.

Biology. Unknown except for the collecting dates (April, May, November).

Distribution. DRC, Ethiopia, Tanzania, Uganda,

Remarks. The holotype female possesses slightly darker head and patagia.

114. *Metarctia (Metarctia) priscilla* Kiriakoff, 1957
(Pl. IV; ♂ Pl. 12)

Short diagnosis. Imago. Forewing length: 14 mm. Forewing brownish black, hindwing paler. Belongs to the group of several dark, externally indistinguishable species.

Male genitalia. Distal process of costa large, wide, triangular with upcurved, sharp tip; ventral protrusion indistinct, densely setose; proximal part of costa convex; saccus elongate, apically rounded.

Female genitalia. Unknown.

Early stages. Unknown.

Biology. Unknown except for the collecting dates (May or June).

Distribution. Ghana.

115. *Metarctia (Metarctia) pulverea* Hampson, 1907
(Pl. IV; ♂ Pl. 12)

Short diagnosis. Imago. Forewing length: 16-20 mm. Differs from all other *Metarctia* by rust brown forewing suffused with numerous black scales with two indistinct dark blotches at end of cell and below base of M_2. Forewing underside paler with one darker blotch at end of cell.

Male genitalia. Uncus relatively short; valva very short and wide; distal process narrow, elongate, longer than valva; portion of juxta surrounding phallus ventrally with row of numerous, irregular teeth; saccus large, triangle shaped.

Female genitalia. Not examined.

Early stages. Unknown.

Biology. Unknown except for the collecting dates (January, September, December).

Distribution. DRC, Rwanda, Uganda.

Remarks. The male genitalia are almost indistinguishable from those of *M. alticola, M. burungae, M. flora*, and *M. virgata*.

116. *Metarctia (Metarctia) pumila* HAMPSON, 1909
(Pl. V; ♂ Pl. 12)

Short diagnosis. Imago. Forewing length: 15-16 mm. Forewing brownish black, hindwing paler. Belongs to the group of several dark, externally indistinguishable species. Most similar to *M. subpallens*.

Male genitalia. Lateral walls of uncus broadened, forming prominent "wings"; distal process of valva short, narrow; ventral protrusion in form of prominent edge terminating at 2/3 length of costa; proximal portion of costa distinctly convex; saccus elongate; phallus narrow, straight.

Female genitalia. Unknown.

Early stages. Unknown.

Biology. Unknown.

Distribution. Sudan.

Remarks. The type series consist of four syntypes (2 ♂♂, 2 ♀♀) which have been found in the BMNH. One of them, a male, with a small, rounded, red-bordered label with "TYPE" was selected and here designated as a LECTOTYPE to prevent any doubts as to the identity of this species. The second male was labeled as PARALECTOTYPE. The two females do not belong to this species, they are in fact the females of *M. phaeoptera*.

117. *Metarctia (Metarctia) robusta* (KIRIAKOFF, 1973)

Short diagnosis. Imago. According to KIRIAKOFF (1973a), forewing length: 23 mm; similar to *M. atrivenata* but much paler.

Male genitalia. Not examined.

Female genitalia. Unknown.

Early stages. Unknown.

Biology. Unknown except for the collecting date (January).

Distribution. Zambia.

Remarks. The type material was not examined. The type should be housed in the BMZ, but according to a personal communication from V. MUYAMBO, it was not located in this collection.

118. *Metarctia (Metarctia) rufescens* WALKER, 1855
(Pl. V; ♂ Pl. 12; ♀ Pl. 30)

Short diagnosis. Imago. Forewing length: 15-20 mm. Most similar to *M. haematricha*. In both species wings uniformly brownish colour but *M. rufescens* can be at once separated by the narrow, dark yellow streak along the subcostal vein. In darker forms the streak may be reduced to an indistinctive suffusion of yellow scales.

Male genitalia. Distal process of valva short; ventral protrusion elongate, narrow, covered with setose margin from base of valva until dorso median area of costa; saccus prominent, elongate.

Female genitalia. Ostium bursae creviced, surrounded by small, sclerotized plate; ductus bursae narrow, membranous, elongate; bursa copulartix heavily plicate, signum absent.

Early stages. The caterpillar and pupa have been described by BARRET (1902). Additional information is provided by TOWNSEND (1944).

Biology. The caterpillar feeds on juicy creeper (*Tradescantia* sp.) and is often found on the ground among dead leaves (BARRET 1902), but according to TOWNSEND (1944) the feeding substrate is "the trash that lies among the roots of grass". According to LE PELLEY (1959) the caterpillar in Kenya feeds on roots of pasture grasses.

Distribution. Burundi, DRC, Kenya, RSA.

Remarks. The record from Kenya is cited from biological data published by LE PELLEY 1959. No specimens from this country were found in collections.

119. *Metarctia (Metarctia) saalfeldi* KIRIAKOFF, 1960
(Pl. V; ♂ Pl. 12)

Short diagnosis. Imago. Forewing length: 12-13 mm. Ochraceous brown species identical to *M. pumila* and *M. subpallens* but distinctly smaller. The forewing length of those species is never less than 15 mm.

Male genitalia. Most similar to *M. fulvia* and *M. pumila*. From the first it can be separated by the reduced ventral protrusion of the valva, the slightly longer and slender distal process, and the more prominent convexity of the proximal part of the costa. The main difference from *M. pumila* is the reduced ventral protrusion of the valva.

Female genitalia. Unknown.

Early stages. Unknown.

Biology. Unknown except for the collecting data (June).

Distribution. Ethiopia.

Remarks. The taxonomic status of this taxon is unclear. Until now only three males are known. Despite many similarities with *M. pumila* and *M. subpallens* the moths are much smaller than all the specimens of these species that I have seen. Additional material is needed to solve the problem of the relationships between these taxa.

The type series of *M. saalfeldi* consists of three males - "holotype et 2 paratypes" (KIRIAKOFF 1960b). All three specimens have the same labels without any indication as to which was selected as holotype. The only useful information is the fact that the genitalia of only one specimen are illustrated. This specimen is regarded as the holotype.

120. *Metarctia (Metarctia) salmonea* KIRIAKOFF, 1957
(Pl. V; ♂ Pl. 12)

Short diagnosis. Imago. Forewing length: 12 mm. Colouration intermediate between that of *M. fario* and *M. pallens*. Wings pinkish cream except for brownish costa of forewing. Head and thorax brownish yellow.

Male genitalia. Distal process of costa elongate, gradually narrowed towards tip; ventral protrusion indistinct, setose; dorsoproximal part of costa with distinct projection; phallus partly damaged.

Female genitalia. Unknown.

Early stages. Unknown.

Biology. Unknown except for the collecting dates (February).

Distribution. Angola.

Remarks. Only one specimen (male) known so far. Taxonomic status still unclear. Possibly a colour form of *M. fario* or *M. pallens*.

121. *Metarctia (Metarctia) sarcosoma* HAMPSON, 1901
(Pl. V; ♂ Pl. 12)

Short diagnosis. Imago. Forewing length: 20 mm. Characterized by orange pink body, brownish black antennae and forewing, and cream cilia and hindwing. Slightly resembling, and confused by KIRIAKOFF (1960), with *M. flavicincta* which is much smaller, delicate, with a brownish thorax.

Male genitalia. Distal process of valva short, wide, ventral protrusion at 2/3 length of valva, indistinct, covered with numerous setae; saccus short, spiculate.

Female genitalia. Unknown.

Early stages. Unknown.

Biology. Unknown except for the collecting dates (March, April, June, December). In Kakamega Forest (Kenya) moths have been collected at an artificial light source located both inside the young secondary forest and in open habitat (farmland).

Distribution. Kenya, Uganda.

Remarks. This species, *M. fontainei*, and *M. flavicincta* form a group of similar taxa for which the taxonomic status is incertain. There are forms with combinations of characters making a proper determination very difficult.

122. *Metarctia (Metarctia) sheljuzhkoi* KIRIAKOFF, 1961
(Pl. V; ♂ Pl. 12)

Short diagnosis. Imago. Forewing length: 14-15 mm. Forewing brownish black, hindwing paler. Belongs to the group of several dark, externally indistinguishable species.

Male genitalia. Distal process of valva very short, upcurved, with sharp tip; dorso-ventral protrusion at 2/3 length of costa; proximal part of costa convex; saccus wide basally then gradually narrowing, sharply tipped.

Female genitalia. Unknown.

Early stages. Unknown.

Biology. Unknown except for the collecting dates (September, October).

Distribution. Ivory Coast.

123. *Metarctia (Metarctia) subpallens* KIRIAKOFF, 1956
(Pl. V; ♂ Pl. 13)

Short diagnosis. Imago. Forewing length: 15 mm. Another pale coloured species similar to *M. burra*. General colouration more yellowish, not pinkish; costa of forewing from underneath dark fuscous.

Male genitalia. Similar to *M. pumila* but uncus without flat "wings" and ventral process of costa less visible.

Female genitalia. Unknown.

Early stages. Unknown.

Biology. Unknown except for the collecting dates (April, May, June).

Distribution. Kenya.

124. *Metarctia (Metarctia) tenera* (KIRIAKOFF, 1973)

Short diagnosis. Imago. According to KIRIAKOFF (1973a), forewing length: 13 mm; similar to *A. pallida* and *M. metaleuca*.

Male genitalia. Not examined.

Female genitalia. Unknown.

Early stages. Unknown.

Biology. Unknown except for the collecting date (February).

Distribution. Zimbabwe.

Remarks. The type (a male) housed in the BMZ (pers. comm. V. Muyambo) is inaccessible for study.

The species may belong to the genus *Automolis* as in the original description (Kiriakoff 1973a) the male genitalia are compared with those of *Automolis meteus* and *A. bicolora*.

125. *Metarctia (Metarctia) transvaalica* (Kiriakoff, 1973)

Short diagnosis. Imago. According to Kiriakoff (1973a), forewing length: 13 mm, colouration orange-brown; hindwing paler.

Male genitalia. Not examined.

Female genitalia. Unknown.

Early stages. Unknown.

Biology. Unknown except for the collecting date (November).

Distribution. RSA.

Remarks. The type (a male) housed in the BMZ (pers. comm. V. Muyambo) is inavailable for study.

According to Kiriakoff (1973a) the unique diagnostic character of this new species is the structure of the genitalia which is different than in any other *Metarctia* species. This, however, should be confirmed after reexamination of the type.

126. *Metarctia (Metarctia) tricolorana* Wichgraf, 1922
(Pl. V; ♂ Pl. 13)

Short diagnosis. Imago. Forewing length: 18 mm. Similar but much larger than *M. phaeoptera*; wings not semi-transparent; cilia and base of hindwing cream; yellow colouration of *M. phaeoptera* replaced with pale rust.

Male genitalia. Distal process of costa elongate but shorter than valva; ventral protrusion at ½ length of costa; saccus short, tapering.

Female genitalia. Unknown.

Early stages. Unknown.

Biology. Unknown except for the collecting dates (March).

Distribution. Uganda.

127. *Metarctia (Metarctia) unicolor* (Oberthür, 1880)
(Pl. V; ♂ Pl. 13)

Short diagnosis. Imago. Forewing length: 23-25 mm. One of the larger species, similar to *M. longipalpis* and *M. atrivenata*, but forewing unicolorous rust brown, paler along costa and at base, slightly semi-transparent behind middle; hindwing paler, pinkish cream, semi-transparent; cilia of both wings yellow.

Male genitalia. Distal process broad, much shorter than valva; proximal portion of costa evenly convex; ventral protrusion weakly marked, densely covered with long, delicate setae; saccus broad basally, triangular; phallus of moderate length, straight, broadening gradually in terminal part.

Female genitalia. Not examined.

Early stages. Unknown.

Biology. Unknown except for the collecting dates (June).

Distribution. Eritrea, Ethiopia.

Remarks. The name of the type locality written on the label attached to the holotype (Fin-Finni) is the forgotten, old name of the present capital of Ethiopia – Addis Ababa. The origin and the meaning of the name "Fin-Fekéré" cited in the original description remains unclear.

The holotype of *M. lateritia aegrota* was not located. Possibly it is still present among Lepidoptera material housed in the MCSNG. However, among the BERIO types loaned to me by R. POGGI the species was not found. It is treated as a synonym of *M. major* by KIRIAKOFF (1960) which is here recognized as a synonym of *M. unicolor*. Another species synonymized with *M. unicolor* is *M. erlangeri*. KIRIAKOFF (1960) included *M. diversa* as a synonym of this species. His interpretation is incorrect as the type specimens, although similar, show differences in colouration. The differences are even more significant in the genitalia, which strongly suggests that they are two distinct species.

128. *Metarctia (Metarctia) uniformis* BETHUNE-BAKER, 1911
(Pl. V)

Short diagnosis. Imago. Forewing length: 15 mm. Belongs to the assemblage of many similarly coloured taxa with a dark forewing and pale (usually cream) hindwing. From them it can be separated at once by the uniformly yellow coloration of the head, thorax, and abdomen.

Male genitalia. Not examined.

Female genitalia. Unknown.

Early stages. Unknown.

Biology. Unknown except for the collecting dates (December).

Distribution. Angola.

129. *Metarctia (Metarctia) upembae* KIRIAKOFF, 1954
(Pl. V; ♂ Pl. 13)

Short diagnosis. Imago. Forewing length: 21 mm. Very characteristic, large species with orange-yellow head, thorax, and abdomen, brown forewing and antennae, and yellowish cream hindwing; cilia of both pairs of wings yellowish cream.

Male genitalia. Distal process short, broad basally, slightly upcurved terminally; proximal portion of costa with distinct convexity; ventral protrusion in form of elongate ridge, densely covered with long setae; sacculus triangular; phallus short, straight.

Female genitalia. Unknown.

Early stages. Unknown.

Biology. Unknown except for the collecting dates (November).

Distribution. DRC.

130. *Metarctia (Metarctia) venustissima* KIRIAKOFF, 1961
(Pl. V; ♂ Pl. 13)

Short diagnosis. Imago. Forewing length: 25 mm. A large *Metarctia* with unique colouration. Head and thorax red, abdomen orange-yellow, forewing brownish black with contrasting red veins, hindwing yellow.

Male genitalia. Basoventral process of valva large, well sclerotized; terminal part of valva short, wide; saccus elongate, narrow, triangular; phallus elongate, narrow, straight.

Female genitalia. Unknown.

Early stages. Unknown.

Biology. Unknown except for the collecting dates (November).

Distribution. DRC.

Remarks. Only the holotype is known until now. Its taxonomic placement is unclear. The morphology of the genitalia suggests an affinity with subgenus *Collocaliodes*. The external characters (pattern and colouration) are typical for members of subgenus *Hebena*. Until the additional material will be analized I follow Kiriakoffs' placement of this species in nominotypical subgenus.

131. *Metarctia (Metarctia) virgata* Joicey & Talbot, 1921
(Pl. V; ♂ Pl. 13; ♀ Pl. 30)

Short diagnosis. Imago. Forewing length: 14-19 mm. Very characteristic species. Forewing brownish black with veins, middle costal area, and outer margin cream; narrow, semi-transparent stripe between veins M_2 and CuA_1; second stripe, perpendicular to inner margin, connecting cell with vein A1-A2.

Male genitalia. Very similar to those of *M. alticola* and *M. flora*; terminal process of costa very narrow and elongate.

Female genitalia. Ostium bursae creviced; antevaginal plate flat, fused with 8th sternite; postvaginal plate in form of upraised ridge; colliculum sclerotized; remaining part of ductus bursae short; membranous, gradually broadened towards corpus bursae.

Early stages. Unknown.

Biology. Unknown except for the collecting dates (March, April, August, September).

Distribution. DRC, Rwanda.

Subgenus *Metarhodia* Kiriakoff, 1953

132. *Metarctia (Metarhodia) confederationis* Kiriakoff, 1961
(Pl. V; ♂ Pl. 13)

Short diagnosis. Imago. Forewing length: 14 mm. Pale, small species superficially resembling *M. heinrichi*; forewing uniformly coloured without marking except for small elongate blotch in basodistal area; hindwing pale pinkish cream.

Male genitalia. Similar to those of other representatives of the subgenus; see **Remarks.** Uncus not gradually narrowing toward tip; lateral margins parallel; terminal hook small; phallus relatively small.

Female genitalia. Unknown.

Early stages. Unknown.

Biology. Unknown except for the collecting dates (February).

Distribution. RSA.

Remarks. The morphology of the male genitalia of all representatives of the subgenus is very similar and generally does not provide any reliable characters enabling unmistakable separation of the species. The available material is too small for dividing interspecific and intraspecific variability. The examination of the vesica did not provide useful characters either.

It is highly probable that an extensive revision based on many specimens from different regions would change the status of some taxa.

133. *Metarctia (Metarhodia) epimela* (KIRIAKOFF, 1979)
(Pl. V; ♂ Pl. 13; ♀ Pl. 30)

Short diagnosis. Imago. Forewing length: 14-19 mm. Another species almost identical with pink form of *M. rubripuncta*; only differs in dark suffusion of distal half of hindwing in male and female.

Male genitalia. Similar to these of other representatives of subgenus; see **Remarks** for *M. confederationis*.

Female genitalia. Indistinguishable from those of *M. heinrichi, M. insignis*, and *M. rubribasa*.

Early stages. Unknown.

Biology. Unknown except for the collecting dates (January, February).

Distribution. Tanzania.

Remarks. Taxonomic status dubious. Most probably a colour form of *M. rubripuncta*.

134. *Metarctia (Metarhodia) heinrichi* KIRIAKOFF, 1961
(Pl. V; ♂ Pl. 13; ♀ Pl. 31)

Short diagnosis. Imago. Forewing length: 15-20 mm. Very similar to *M. rubripuncta*; can be separated by paler forewing colour, with more olive tinge and small pink blotches in dorsobasal area and at distal end of cell; hindwing pale pink, cilia concolorous.

Male genitalia. Similar to those of other representatives of subgenus; see **Remarks** for *M. confederationis*.

Female genitalia. Indistinguishable from those of *M. epimela, M. insignis*, and *M. rubribasa*.

Early stages. Unknown.

Biology. Unknown except for the collecting dates (October).

Distribution. Angola, DRC (south part).

135. *Metarctia (Metarhodia) heringi* KIRIAKOFF, 1957
(Pl. V; ♂ Pl. 14; ♀ Pl. 31)

Short diagnosis. Imago. Forewing length: 13-14 mm. Very much like *M. hypomela*, but smaller and with hindwing paler.

Male genitalia. Similar to those of other representatives of subgenus; see **Remarks** for *M. confederationis*.

Female genitalia. Colliculum membranous followed by narrow sclerotized ring.

Early stages. Unknown.

Biology. Unknown except for the collecting dates (August, November, December).

Distribution. Botswana, DRC.

Remarks. Female genitalia significantly different than in other species of subgenus.

136. *Metarctia (Metarhodia) hypomela* KIRIAKOFF, 1956
(Pl. V; ♂ Pl. 14)

Short diagnosis. Imago. Forewing length: 15-17 mm. Resembling small *M. rubripuncta*, but pale colouration rather pinkish or olive cream not pink. Forewing dark ochraceous with pale markings much reduced.

Male genitalia. Similar to those of other representatives of subgenus; see **Remarks** for *M. confederationis*.

Female genitalia. Unknown.

Early stages. Unknown.

Biology. Unknown. Adults collected throughout the year. In Kakamega Forest (Kenya) moths have been collected at an artificial light source located both inside the middle-aged and young secondary forest and in open habitats (farmland, grassland).

Distribution. Kenya.

Remarks. More material is needed for the proper interpretation of the taxonomic position of this species. A paratype male is labeled "Mt. Elgon, Kenya; Sept 1951; T.H.E. Jackson". [not April 1953 as presented in the original description].

137. *Metarctia (Metarhodia) insignis* KIRIAKOFF, 1959
(Pl. V; ♂ Pl. 14; ♀ Pl. 31)

Short diagnosis. Imago. Forewing length: 18-20 mm. Very much like *M. rubripuncta*. Relatively large with wide forewing and bright pink hindwing (especially in its basal part); with pink spot at end of forewing cell.

Male genitalia. Similar to those of other representatives of subgenus; see **Remarks** for *M. confederationis*.

Female genitalia. Indistinguishable from those of *M. epimela*, *M. heinrichi*, and *M. rubribasa*.

Early stages. Unknown.

Biology. Unknown except for the collecting dates (April).

Distribution. Rwanda.

Remarks. Described from one male and one female. Belongs to the group of taxa that are very similar to *M. rubripuncta*. Due to the lack of material and small differences between species, their proper interpretation is currently impossible.

138. *Metarctia (Metarhodia) jordani* KIRIAKOFF, 1957
(Pl. V; ♂ Pl. 14)

Short diagnosis. Imago. Forewing length: 12 mm. Similar in colouration to *M. heinrichi* but slightly darker, with more prominent pink markings on forewing: in dorsobasal area, along costa, in distal part of cell, and between veins R_5 and M_3 near their base; cilia always dark brown.

Male genitalia. Middle process of valva extremely long, more than two times longer than costal process (in holotype).

Female genitalia. Unknown.

Early stages. Unknown.

Biology. Unknown except for the collecting dates (March).

Distribution. Angola.

Remarks. Only two males are known.

139. *Metarctia (Metarhodia) nigricornis* Debauche, 1942
(Pl. V; ♂ Pl. 14)

Short diagnosis. Imago. Forewing length: 20 mm. Very characteristic species, uniformly rust coloured; veins on forewing darker.

Male genitalia. Similar to those of other representatives of subgenus; see **Remarks** for *M. confederationis*.

Female genitalia. Unknown.

Early stages. Unknown.

Biology. Unknown except for the collecting dates (June).

Distribution. DRC.

140. *Metarctia (Metarhodia) rubribasa* Bethune-Baker, 1911
(Pl. V; ♂ Pl. 14; ♀ Pl. 31)

Short diagnosis. Imago. Forewing length: 17-23 mm. The only species in the subgenus with such a pale rust-brown forewing and much paler pinkishcream hindwing; forewing at base, cell, and costa suffused with pink, but this varies between individuals.

Male genitalia. Similar to those of other representatives of subgenus; see **Remarks** for *M. confederationis*.

Female genitalia. Indistinguishable from those of *M. epimela*, *M. heinrichi*, and *M. insignis*.

Early stages. Unknown.

Biology. Unknown except for the collecting dates (March, October-December).

Distribution. Angola, DRC, Tanzania.

Remarks. The species was described from a single female. *Metarctia (M.) deriemaeckeri* Kiriakoff, 1953 is here synonymized with *M. rubribasa*.

141. *Metarctia (Metarhodia) rubripuncta* Hampson, 1898
(Pl. V; ♂ Pl. 14; ♀ Pl. 32)

Short diagnosis. Imago. Forewing length: 14-26 mm. Very variable in size and colouration. Forewing dark, with small, pink blotch at base and distal end of the cell. Hindwing concolorous with forewing (typical form) or pink, but extremely variable. Forewing markings can be reduced, or almost entirely absent.

Male genitalia. Similar to those of other representatives of subgenus; see **Remarks** for *M. confederationis*.

Female genitalia. Ductus bursae narrow and elongate.

Early stages. Unknown.

Biology. *Ipomea batatas* was recorded as a host plant in Kenya (Le Pelley 1959). Adults collected throughout the year. In Kakamega Forest (Kenya) moths have been collected at an artificial light source located inside the middle-aged and young secondary forest.

Distribution. Burundi, Cameroon, DRC, Gabon, Kenya.

Remarks. Very variable species. Some of the similar species described by Kiriakoff may in fact belong to this taxon. The holotype of *M. impura*, which is a synonym of *M. rubripuncta*, was collected on 13.11.1957, not 13.02.1952 as cited incorrectly in the original description.

Subgenus *Pinheyata* NYE in WATSON, FLETCHER & NYE, 1995

142. *Metarctia (Pinheyata) crocina* (KIRIAKOFF, 1973)

Short diagnosis. Imago. According to KIRIAKOFF (1973a), forewing length: 16-18 mm; colouration as in *M. rufescens*, therefore confused with this species.

Male genitalia. Not examined, but according to KIRIAKOFF (1973a) resembling those of *Owambarctia*.

Female genitalia. Unknown.

Early stages. Unknown.

Biology. Unknown except for the collecting dates (February, October).

Distribution. Zimbabwe.

Remarks. The type housed in the BMZ (pers. comm. V. MUYAMBO) is inaccessible for study. Known only from the type series of an unknown number of males (KIRIAKOFF 1973a).

143. *Metarctia (Pinheyata) quinta* (KIRIAKOFF, 1973)
(Pl. V; ♂ Pl. 14)

Short diagnosis. Imago. Forewing length: 15 mm. Forewing pale fuscous, darker near base; hindwing orange-cream, unicolorous; patagia, tegula, and costa of forewing pale orange.

Male genitalia. Costa of valva terminated with sclerotized, elongate process; inner margin of costa covered with long, dense setae.

Female genitalia. Unknown.

Early stages. Unknown.

Biology. Unknown except for the collecting dates (January, February, May, June, October, November).

Distribution. Zimbabwe.

Remarks. The type housed in the BMZ (pers. comm. V. MUYAMBO) is inaccessible for study. The photograph of imago and male genitalia were taken from specimens determined by KIRIAKOFF and housed in the RMCA.

Subgenus *Thyretarctia* STRAND, 1912

144. *Metarctia (Thyretarctia) brunneoaurantiaca* (KIRIAKOFF, 1973)

Short diagnosis. Imago. Forewing length: 11 mm. Known to me only from the literature. According to the original description, much paler than any of the *Thyretarctia* species.

Male genitalia. Not examined. According to KIRIAKOFF (1973a), similar to those of *M. schoutedeni*, but phallus is "unbewaffnet" [possibly without cornuti] and "Fortsätze der Valva sind aber etwas mehr verlängert" [terminal processes of valva are slightly more elongate].

Female genitalia. Unknown.

Early stages. Unknown.

Biology. Unknown except for the collecting date (April).

Distribution. Kenya.

Remarks. The holotype, housed in the BMZ (pers. comm. V. MUYAMBO) is the only known specimen and is inaccessible for study.

145. *Metarctia (Thyretarctia) didyma* Kiriakoff, 1957
(Pl. V; ♂ Pl. 14)

Short diagnosis. Imago. Forewing length: 10-13 mm. Externally identical to *M. haematica*. **Male genitalia.** Both processes of valva pointed and of equal length as opposed to *M. haematica* in which costal process is longer and narrower.

Female genitalia. Not examined.

Early stages. Unknown.

Biology. Unknown.

Distribution. Chad, DRC, Ghana, Ivory Coast, Niger.

Remarks. Possibly a synonym of *M. haematica*.

146. *Metarctia (Thyretarctia) haematica* Holland, 1893
(Pl. V; ♂ Pl. 15; ♀ Pl. 32)

Short diagnosis. Imago. Forewing length: 12-16 mm. Indistinguishable from *M. didyma* by external features except for the bigger size.

Male genitalia. Membranous pouch of vesica terminated with prominent, sclerotized tooth more than three times as long as wide; group of numerous cornuti forming narrow, elongate protrusion.

Female genitalia. Corpus bursae large, rounded, with prominent membranous process in posterior part; signum rounded, weekly sclerotized.

Early stages. Unknown.

Biology. The caterpillar feeds on *Aster* sp. (Asteraceae) (Sevastopulo 1975). Adults collected throughout the year. In Kakamega Forest (Kenya) moths have been collected at an artificial light source located mainly in open habitat (farmland) but ilso inside the middle-aged and young secondary forest.

Distribution. DRC, Gabon, Kenya.

Remarks. In the CMNH there are two males of *M. haematica* bearing similar labels with handwritten inscription "TYPE". However the specimens are different in size and only the bigger one matches the size given by Holland in his description ("expanse, 28 mm"). Therefore I consider it as the HOLOTYPE while the second, smaller specimen does not belong to the type series. The wingspan of this smaller specimen is 24,5 mm.

147. *Metarctia (Thyretarctia) infausta* Kiriakoff, 1957
(Pl. V; ♂ Pl. 15)

Short diagnosis. Imago. Forewing length: 14 mm. Similar to *M. schoutedeni* with which it forms a pair of dark coloured species. However, it is darker with a slightly paler forewing basal half, without any trace of brownish tinge.

Male genitalia. Almost indistinguishable from those of *M. schoutedeni*; group of cornuti slightly shorter.

Female genitalia. Unknown.

Early stages. Unknown.

Biology. Unknown except for the collecting dates (March).

Distribution. DRC.

Remarks. Known from two males from Kibali-Ituri: Nioka. Possibly only a colour variety of *M. schoutedeni*, but this should be studied on extensive material.

148. *Metarctia (Thyretarctia) morosa* KIRIAKOFF, 1957
(Pl. V; ♂ Pl. 15)

Short diagnosis. Imago. Forewing length: 14 mm. Uniformly dark ochraceous species with cream antennae.

Male genitalia. Indistinguishable from those of *M. schoutedeni* and *M. haematica*.

Female genitalia. Unknown.

Early stages. Unknown.

Biology. Unknown except for the collecting dates (March).

Distribution. DRC (Elisabethville).

Remarks. Phallus lost. It was not found in the plastic vial pinned near the specimen when the genitalia were mounted on slide.

Known from the very worn male holotype.

149. *Metarctia (Thyretarctia) schoutedeni* KIRIAKOFF, 1953
(Pl. V; ♂ Pl. 15; ♀ Pl. 32)

Short diagnosis. Imago. Forewing length: 13-22 mm. Similar to *M. infausta* but with more or less prominent brown or rust brown tinge covering the dorsobasal half of the forewing.

Male genitalia. Membranous pouch of vesica terminated with prominent, sclerotized tooth about two times as long as wide; group of numerous cornuti forming small, rounded protrusion.

Female genitalia. Corpus bursae slightly elongate, without prominent membranous process; signum elongate, heavily sclerotized.

Early stages. Unknown.

Biology. Unknown except for the collecting dates (January-May, September-December). In Kakamega Forest (Kenya) moths have been collected at an artificial light source located both inside the middle-aged and young secondary forest and in open habitat (farmland).

Distribution. DRC, Kenya.

Remarks. *Metarctia (T.) pinheyi* KIRIAKOFF, 1956 was synonymized with *M. schoutedeni* by PRZYBYŁOWICZ & KÜHNE 2008: 152. Neither genitalia nor colouration provide characters for separating them.

Microbergeria KIRIAKOFF, 1972

150. *Microbergeria luctuosa* KIRIAKOFF, 1972

Short diagnosis. Imago. Forewing length: 12 mm. No specimens were examined.

Male genitalia. Not examined.

Female genitalia. Unknown.

Early stages. Unknown.

Biology. Unknown except for the collecting dates (September).

Distribution. Cameroon.

Remarks. The type housed in the BMZ (pers. comm. V. MUYAMBO) is inaccessible for study. The genitalia of the holotype illustrated in the original description look very similar to those of *Pseudothyretes nigrita*.

Neophemula Kiriakoff, 1957

151. *Neophemula vitrina* (Oberthür, 1909)
(Pl. V; ♂ Pl. 15)

Short diagnosis. Imago. Forewing length: 12-15 mm. Differs from all other Thyretini by the greyish white colouration of the body, the brown, iridescent, basal and the semi-transparent central parts of the forewing, and the extremely reduced hindwing.

Male genitalia. Valva sclerotized, costal process large, flattened posteriorly, slightly concave terminally; sacculus elongated, pointed apically, setose distally.

Female genitalia. Unknown.

Early stages. Unknown.

Biology. Unknown except for the collecting dates (February-May, September, October).

Distribution. Angola, Cameroon, DRC, Gabon, Guinea.

Remarks. The taxonomic status of subspecies *congoensis* Kiriakoff, 1957a and *angolensis* Kiriakoff, 1957c remains unclear. So far only 10 specimens of *N. vitrina* are known.

Owambarctia Kiriakoff, 1957

152. *Owambarctia owamboensis* Kiriakoff, 1957
(Pl. V; ♂ Pl. 15)

Short diagnosis. Imago. Forewing length: 13-14 mm. Body unicolorous, dark brown, hindwing slightly paler than cilia. Differs from *O. unipuncta* only by the shape of genitalia.

Male genitalia. Valva elongate, costa slightly concave, apex of valva rounded, sacculus slightly convex with the sharp sclerotized process rounded on the tip with some hairs on it.

Female genitalia. Unknown.

Early stages. Unknown.

Biology. Unknown.

Distribution. Namibia.

Remarks. Known from the type series only. The paratype described as a female is in fact a male. Female remains unknown. The taxonomic status of both *Owambarctia* species is unresolved.

153. *Owambarctia unipuncta* Kiriakoff, 1973
(Pl. VI; ♂ Pl. 15)

Short diagnosis. Imago. Forewing length: 14-15 mm. As *O. owamboensis*, but holotype paler.

Male genitalia. Differs from *O. owamboensis* in more triangular shape of valva terminating with prominent, sclerotized, pointed, process; basal part of costa with prominent lobe densely covered ventrally with, elongate setae.

Female genitalia. Unknown.

Early stages. Unknown.

Biology. Unknown except for the collecting dates (January, March, August, December).

Distribution. Tanzania.

Remarks. Female unknown. Taxonomic status unclear.

Paramelisa Aurivillius, 1905

154. *Paramelisa dollmani* Hampson, 1920
(Pl. VI; ♂ Pl. 15)

Short diagnosis. Imago. Forewing length: 17-20 mm. Patagia and tuft of long hair at apex of abdomen pale yellow. Narrow transverse bands on abdominal tergites crimson.

Male genitalia. Very similar to those of *P. lophura* but uncus narrower, ventral forks longer; small appendix of terminal part of phallus weakly developed.

Female genitalia. Not examined.

Early stages. Unknown.

Biology. Unknown except for the collecting dates (February, August-November).

Distribution. DRC, Uganda, Zambia.

Remarks. The female of this species, known until now only from one specimen, was not dissected to preserve important characters located on the abdominal terga.

The male, with a small rounded label "TYPE" was selected and here designated as the LECTOTYPE to prevent any doubts as to the identity of this species. The second specimen (female) was labeled as the PARALECTOTYPE.

155. *Paramelisa leroyi* Kiriakoff, 1953
(Pl. VI)

Short diagnosis. Imago. Forewing length: 16 mm. Patagia pale yellow; tuft of long hair at apex of abdomen brown.

Male genitalia. Not examined.

Female genitalia. Unknown.

Early stages. Unknown.

Biology. Unknown except for the collecting dates (February).

Distribution. DRC.

Remarks. The species is known only from one male. In this genus many important characters (pattern, colours) are located on the abdomen and to avoid their destruction the genital organs were not prepared.

156. *Paramelisa lophura* Aurivillius, 1905
(Pl. VI; ♂ Pl. 15; ♀ Pl. 32)

Short diagnosis. Imago. Forewing length: 18-26 mm. The only species without red or pink markings on the thorax and abdomen.

Male genitalia. As *P. dollmani* except for broader uncus, longer ventral forks, and prominent appendix of terminal part of phallus.

Female genitalia. Ventral pheromone glands sack-like; ductus bursae narrow, elongate, membranous; corpus bursae prominent, spherical with small, rounded, sclerotized signum in its central part.

Early stages. Pupa described by AURIVILLIUS (1905b): dark brown, in sparse, soft cocoon interspaced with numerous, short, stiff hairs.

Biology. Unknown except for the collecting dates (January, March, July, September, October, December). In Kakamega Forest (Kenya) moths have been collected at an artificial light source located inside the middle-aged and young secondary forest.

Distribution. DRC, Gabon, Kenya, Uganda.

157. *Paramelisa lophuroides* OBERTHÜR, 1911
(Pl. VI)

Short diagnosis. Imago. Forewing length: 17 mm. Patagia and tuft of long hair at apex of abdomen brown, not pale yellow.

Male genitalia. Not examined.

Female genitalia. Unknown.

Early stages. Unknown.

Biology. Unknown except for the collecting dates (May).

Distribution. Cameroon.

Remarks. The species is known only from one male which was not dissected to avoid the loss of important characters located on the abdomen.

Pseudmelisa HAMPSON, 1910

158. *Pseudmelisa chalybsa* HAMPSON, 1910
(Pl. VI; ♀ Pl. 33)

Short diagnosis. Imago. Forewing length: 18-19 mm. Base of tegula red; apical part of abdomen yellow red.

Male genitalia. Unknown.

Female genitalia. Ductus bursae narrow, 5-6 times longer than wide.

Early stages. Unknown.

Biology. Unknown except for the collecting dates (October, November).

Distribution. DRC, Zambia.

Remarks. The male of this species most probably has been described in genus *Metamicroptera*.

159. *Pseudmelisa rubrosignata* KIRIAKOFF, 1957
(Pl. VI; ♀ Pl. 33)

Short diagnosis. Imago. Forewing length: 18 mm. Base of tegula and apical part of abdomen black.

Male genitalia. Unknown.

Female genitalia. Ductus bursae very narrow, more than six times longer than wide.

Early stages. Unknown.

Biology. Unknown except for the collecting dates (February).

Distribution. Malawi, Tanzania.

Remarks. Only three females known. It may be a colour form of *P. chalybsa*, but this problem requires more material for study.

Pseudothyretes Dufrane, 1945

160. *Pseudothyretes carnea* (Hampson, 1898)
(Pl. VI)

Short diagnosis. Imago. Forewing length: 22 mm. Body rust brown, wings semi-transparent; forewing with two rows of transparent spots; hindwing with one transparent spot beyond cell.

Male genitalia. Unknown.

Female genitalia. Unknown. The distal end of the abdomen is lost on the holotype.

Early stages. Unknown.

Biology. Unknown.

Distribution. Angola.

Remarks. One of the two species in the genus described from a female. It certainly is conspecific with one of the other species but it is impossible now to match females with males. All known females are externally similar. Genital differences between females do not allow for proper ascription to known male. The finding of copulating pairs or rearings could clarify these taxonomic problems.

161. *Pseudothyretes erubescens* (Hampson, 1901)
(Pl. VI; ♂ Pl. 16)

Short diagnosis. Imago. Forewing length: 12-16 mm. Comparatively large species; forewing semi-transparent, deep brown with rather small hyaline blotches; hindwing distinctly pinkish brown with small transparent blotch medially.

Male genitalia. Lobes of uncus moderately broadened terminally.

Female genitalia. Unknown.

Early stages. Unknown.

Biology. Unknown except for the collecting dates (April, November).

Distribution. DRC, Kenya, Rwanda, Uganda.

Remarks. The holotype bears no abdomen.

162. *Pseudothyretes kamitugensis* (Dufrane, 1945)
(Pl. VI; ♂ Pl. 16)

Short diagnosis. Imago. Forewing length: 12-15 mm. Another large species; forewing pale rust ochraceous, outer area distinctly paler, semi-transparent blotches large, basal blotch always visible; hindwing ochraceous cream, semi-transparent.

Male genitalia. Uncus unusual for the genus, without long parallel lobes, terminated dorsally by two small processes.

93

Female genitalia. Unknown.

Early stages. Unknown.

Biology. Unknown except for the collecting dates (March, August-October). In Kakamega Forest (Kenya) moths have been collected at an artificial light source located inside the middle-aged and young secondary forest.

Distribution. DRC, Kenya, Rwanda.

163. *Pseudothyretes mariae* DUFRANE, 1945

Short diagnosis. Imago. Forewing length: 20 mm. As in *P. carnea.*

Male genitalia. Unknown.

Female genitalia. Not examined.

Early stages. Unknown.

Biology. Unknown.

Distribution. DRC.

Remarks. The holotype is in fact a female, not a male as mentioned by DUFRANE (1945). For taxonomic interpretation see **Remarks** for *P. carnea.*

164. *Pseudothyretes nigrita* (KIRIAKOFF, 1961)
(Pl. VI; ♂ Pl. 16)

Short diagnosis. Imago. Forewing length: 11-13 mm. Dark, small species with slightly paler hindwing and small hyaline blotches.

Male genitalia. Lobes of uncus elongate, rounded, each provided with small pointed process medially.

Female genitalia. Unknown.

Early stages. Unknown.

Biology. Unknown except for the collecting dates (January, July-September). In Kakamega Forest (Kenya) moths have been collected at an artificial light source located inside the forest.

Distribution. DRC.

Remarks. The holotype is an extremely dark individual. For an accurate determination the examination of the genitalia is required.

165. *Pseudothyretes perpusilla* (WALKER, 1856)
(Pl. VI; ♂ Pl. 16)

Short diagnosis. Imago. Forewing length: 11-14 mm. Indistinguishable from *P. nigrita.*

Male genitalia. Lobes of uncus long, narrow, slightly widened towards extremity.

Female genitalia. Unknown.

Early stages. Unknown.

Biology. Adults collected throughout the year (except July). Information on nocturnal activity in Uganda is provided by TJÖNNELAND (1962). The species was collected at light mostly before midnight but no earlier than about one hour after sunset and only singly during the rest of the time. Specimens were never collected at dusk or early morning. In Kakamega Forest (Kenya) moths have been collected at an artificial light source located both inside the middle-aged and young secondary forest and in open habitat (farmland).

Distribution. Kenya.

Remarks. Easy to separate from other *Pseudothyretes* species on the basis of genitalic characters.

166. *Pseudothyretes rubicundula* (Strand, 1912)
(Pl. VI; ♂ Pl. 16)

Short diagnosis. Imago. Forewing length: 11-14 mm. Size and colouration very similar to those of *P. nigrita* and *P. perpusilla.*

Male genitalia. Lobes of uncus short, apically very wide.

Female genitalia. Unknown.

Early stages. Unknown.

Biology. Unknown except for the collecting dates (May, November).

Distribution. DRC, Equatorial Guinea, Kenya.

Remarks. The variability in morphology of the male genitalia of *P. nigrita, P. perpusilla* and *P. rubicundula* should be analyzed on a larger material from a wide area of Equatorial Africa.

Rhabdomarctia Kiriakoff, 1953

167. *Rhabdomarctia rubrilineata* (Bethune-Baker, 1911)
(Pl. VI; ♂ Pl. 16; ♀ Pl. 33)

Short diagnosis. Imago. Forewing length: 15-24 mm. Body brown to dark brown with red to yellow (rare form) markings on forewing; markings often reduced, especially in females.

Male genitalia. Valva triangular with group of stout, long setae in terminal area.

Female genitalia. Ductus bursae narrow, elongate, slightly curved; colliculum strongly sclerotized; corpus bursae without signum.

Early stages. Unknown.

Biology. Unknown. Adults collected throughout the year. In Kakamega Forest (Kenya) moths have been collected at an artificial light source located both inside the middle-aged and young secondary forest and in open habitats (farmland, grassland).

Distribution. Angola, Cameroon, DRC, Kenya, Tanzania, Uganda.

Remarks. Several species described by Kiriakoff were synonymized as individual colour forms of this relatively variable species. The genitalia do not show any significant differences and well match the internal variability observed in *R. rubrilineata.*

Rhipidarctia Kiriakoff, 1953

Subgenus *Elsitia* Przybyłowicz [nom. nov.]

168. *Rhipidarctia (Elsitia) cinctella* (Kiriakoff, 1953)
(Pl. VI; ♂ Pl. 16)

Short diagnosis. Imago. Forewing length: 17-21 mm. Very similar to *R. saturata*, but male and female with forewing unicolorously dark ochraceous except for red, narrow line along costa of male. Female abdomen with narrow yellow or red fascia on distal margin of each tergite.

Male genitalia. Indistinguishable from *R. saturata.*

Female genitalia. Not examined.

Early stages. Unknown.

Biology. Unknown except for the collecting dates (February-April, June, August-December).

Distribution. DRC.

Remarks. The type series consist of two females (holotype and paratype). In original description ten additional specimens (all females) are mentioned although they are explicitly excluded from the type series. *Rhipidarctia strenua* Kiriakoff, 1957 is here regarded as a synonym. The type series of *R. strenua* consists of a male and a female with reduced (hardly visible) yellow fasciae on the abdomen. In the original description *R. strenua* was compared with *R. forsteri* and *R. unicolor* (now synonym of *R. forsteri*) but it differs by the bigger size and absence of cream markings on the forewing. The holotype of *R. strenua* was collected 13.11.1952, not 13.09.1952 as incorrectly cited in the original description.

169. *Rhipidarctia (Elsitia) forsteri* (Kiriakoff, 1953)
(Pl. VI; ♂ Pl. 16; ♀ Pl. 33)

Short diagnosis. Imago. Forewing length: 15-17 mm. The only *Rhipidarctia* species with uniformly brownish black forewing provided with a small cream dot in the cell. Most specimens bear additional dots beyond cell and between veins CuA_2 and A1-A2.

Male genitalia. Saccus large, elongate (this character is clearly visible in the holotypes of *R. forsteri* and *R. punctulata* (a syn. of *R. forsteri*)).

Female genitalia. Membranous zone posterior from ostium bursae very broad, widely open terminally; 8th sternite more thickly sclerotized than 7th sternite.

Early stages. Unknown.

Biology. Unknown. Adults collected throughout the year. In Kakamega Forest (Kenya) moths have been collected at an artificial light source located mainly inside the middle-aged forest.

Distribution. DRC, Kenya, Rwanda, Uganda.

170. *Rhipidarctia (Elsitia) lutea* (Holland, 1893)
(Pl. VI; ♂ Pl. 16; ♀ Pl. 34)

Short diagnosis. Imago. Forewing length: 13-16 mm. Similar to *R. subminiata* but yellow.

Male genitalia. Valva narrow, rounded terminally, the length of the half of pseudovalva, ventrodistally with long setae.

Female genitalia. 7th sternite laterally and anteriorly without setae; membranous zone posterior from ostium bursae with lateral margins parallel; anterior apophyses short, membranous, curved.

Early stages. Unknown.

Biology. Unknown except for the collecting dates (January-October).

Distribution. DRC.

Remarks. Most probably only a colour form of *R. subminiata* or *R. pareclecta*.

171. *Rhipidarctia (Elsitia) pareclecta* (HOLLAND, 1893)
(Pl. VI; ♂ Pl. 17; ♀ Pl. 34)

Short diagnosis. Imago. Forewing length: 14-17 mm. Similar to *R. crameri* but forewing always with dark suffusion in middle part of interspace between M2 and A1-A2.

Male genitalia. Valva larger, much better developed than weakly sclerotized pseudovalva.

Female genitalia. As *R. rubrovitta* but setae on 8th abdominal sternite more numerous; anterior apophyses slightly shorter, membranous.

Early stages. Unknown.

Biology. Unknown. Adults collected throughout the year. In Kakamega Forest (Kenya) moths have been collected at an artificial light source located mainly inside the middle-aged and young secondary forest but also less numerously in open habitat (farmland).

Distribution. Angola, Gabon, Kenya, Uganda.

172. *Rhipidarctia (Elsitia) saturata* KIRIAKOFF, 1957
(Pl. VI; ♂ Pl. 17)

Short diagnosis. Imago. Forewing length: 16 mm. Resembling *R. cinctella* but always with dark markings as found in *R. pareclecta*.

Male genitalia. Indistinguishable from *R. cinctella*.

Female genitalia. Unknown.

Early stages. Unknown.

Biology. Unknown except for the collecting dates (January, December).

Distribution. DRC.

173. *Rhipidarctia (Elsitia) subminiata* KIRIAKOFF, 1959
(Pl. VI; ♂ Pl. 17)

Short diagnosis. Imago. Forewing length: 14-17 mm. Very similar to *R. crameri*, but colouration orange pink; mainly differs in absence of extensive dark suffusion in basal half of interspace between CuA_1 and CuA_2.

Male genitalia. Valva narrow, rounded terminally, shorter than half of length of pseudovalva.

Female genitalia. Unknown.

Early stages. Unknown.

Biology. Unknown except for the collecting dates (December).

Distribution. DRC.

Remarks. In contrast to other yellow or orange *Rhipidarctia* species, the individual variation in overall colouration is very little.

Subgenus *Hemirhipidia* KIRIAKOFF, 1955

174. *Rhipidarctia (Hemirhipidia) postrosea* (ROTHSCHILD, 1913)
(Pl. VI; ♂ Pl. 17; ♀ Pl. 34)

Short diagnosis. Imago. Forewing length: 13-20 mm. The only *Rhipidarctia* with a yellow forewing provided with brownish markings, and with the hindwing cream-pink to pink.

Male genitalia. Pseudovalva widened terminally, dorso-terminal portion distinctly elongate forming sclerotized process.

Female genitalia. As in *R. aurora*.

Early stages. Unknown.

Biology. Unknown except for the collecting dates (June, July, October, November).

Distribution. Cameroon, DRC, Nigeria.

Subgenus *Rhipidarctia* KIRIAKOFF, 1953

175. *Rhipidarctia (Rhipidarctia) aurora* KIRIAKOFF, 1957
(Pl. VI; ♂ Pl. 17; ♀ Pl. 34)

Short diagnosis. Imago. Forewing length: 11-15 mm. Together with *R. flaviceps* and *R. rhodosoma* it forms a group of similarly dark coloured species. From both of them it can be separated by the uniformly blackish brown colouration of the forewing and rust head, thorax, and abdomen; hindwing concolorous with forewing except dark reddish brown base and anal part.

Male genitalia. Indistinguishable from those of the other species. See **Remarks**.

Female genitalia. Anterior apophyses membranous, extremely short, two times as long as wide; ostium bursae located in rhomboidal depression with distal wall, plicate; indistinguishable from those of *R. postrosea*.

Early stages. Unknown.

Biology. Unknown except for the collecting dates (April, August).

Distribution. DRC.

Remarks. The male holotype lacks of most of the scales on the thorax and abdomen, and the left pair of wings is entirely semi-transparent (descaled).

The male genitalia of all representatives of the subgenus (except *R. crameri*) are so similar that they do not allow for species differentiation. A study based on extensive material should be undertaken to judge the taxonomic value of the described species. The small differences that can be observed in the genitalia of the illustrated specimens (in the shape and length of the uncus and valva) shall be treated with reservation in the process of species determination. To avoid any subjective interpretation, the genitalia of the types are presented when possible.

176. *Rhipidarctia (Rhipidarctia) conradti* OBERTHÜR, 1911
(Pl. VI; ♂ Pl. 17; ♀ Pl. 35)

Short diagnosis. Imago. Forewing length: 11-14 mm. Very similar to *R. crameri* from which it can by separated by the orange pink, not yellow or orange, colouration of the body. Hindwing underneath with dark brown terminal part of costa and veins, and with concolorous cilia.

Male genitalia. Indistinguishable from those of the other species. See **Remarks** for *R. aurora*.

Female genitalia. Indistinguishable from those of *R. crameri.*

Early stages. Unknown.

Biology. Unknown except for the collecting dates (October).

Distribution. Cameroon, DRC.

Remarks. Possibly it is only an individual colour form of *R. crameri.*

177. *Rhipidarctia (Rhipidarctia) crameri* KIRIAKOFF, 1961
(Pl. VI; ♂ Pl. 17; ♀ Pl. 35)

Short diagnosis. Imago. Forewing length: 13-17 mm. Very variable in terms of colouration and markings, but hindwing pale yellow to dark orange, always without any trace of black scales. The darkest forms may be confused with *R. conradti* which, however, is more pinkish.

Male genitalia. Valva relatively long and wide, tip almost reaching the outer margin of pseudovalva; outer zone of spinules of pseudovalva very wide, covering approximately half of its length.

Female genitalia. Anterior apophyses very short; membranous zone posterior from ostium bursae very narrow, proximal part broader than terminal.

Early stages. Unknown.

Biology. Unknown except for the collecting dates (June, September-November). In Kakamega Forest (Kenya) moths have been collected at an artificial light source located both inside the middle-aged and young secondary forest and in open habitat (farmland).

Distribution. DRC, Kenya, Rwanda, Uganda.

Remarks. Very common species recorded from many countries in Equatorial Africa. Most faunistic data concerning *R. invaria* in fact refer to this species.

178. *Rhipidarctia (Rhipidarctia) flaviceps* (HAMPSON, 1898)
(Pl. VI; ♂ Pl. 17; ♀ Pl. 35)

Short diagnosis. Imago. Forewing length: 10-15 mm. It can be separated at once from *R. aurora* and *R. rhodosoma* by the brownish black colouration of the thorax and abdomen.

Male genitalia. Indistinguishable from those of the other species. See **Remarks** for *R. aurora.*

Female genitalia. As in *R. invaria.*

Early stages. Unknown.

Biology. Unknown except for the collecting dates (January-March, June, September, October, December).

Distribution. Cameroon, DRC, Equatorial Guinea (only Bioko Island).

Remarks. The validity of this taxon is uncertain. It is very similar to *R. invaria* and the variation within and between these two taxa should be analyzed with extensive material.

179. *Rhipidarctia (Rhipidarctia) invaria* (WALKER, 1856)
(Pl. VI; ♂ Pl. 17; ♀ Pl. 35)

Short diagnosis. Imago. Forewing length: 13-15 mm. Differs from all other *Rhipidarctia* by the black hindwing veins and cilia. The only species with similar markings on the hindwing is *R. miniata* which, however, is rust-red and without markings on the forewing except for black veins.

Male genitalia. Indistinguishable from those of the other species. See **Remarks** for *R. aurora*.

Female genitalia. Setae concentrated in median portion of 7th sternite; with pair of sclerotized, distally directed protrusions anterolateral from ostium bursae; ostium bursae separated from narrow, membranous zone by small, plicate, sclerotized tooth.

Early stages. Information on the morphology of the caterpillar and pupa has been provided by STRAND (1912) and ALIBERTI (1951).

Biology. The caterpillar was found on cacao *Theobroma cacao* (Sterculiaceae) (ALIBERTI 1951), *Aster* sp., and *Zinnia* sp. (Asteraceae) (SEVASTOPULO 1975). Moths were collected in March, May and September.

Distribution. DRC, Gabon, Ivory Coast, Nigeria, Sierra Leone.

Remarks. For a long time the species was confused with *R. crameri*. Most of the published faunistic data probably concern the later species.

180. *Rhipidarctia (Rhipidarctia) miniata* KIRIAKOFF, 1957
(Pl. VI; ♀ Pl. 36)

Short diagnosis. Imago. Forewing length: 13 mm. Differs from all other *Rhipidarctia* species by the uniformly rust-red fore and hindwing with black veins and cilia, without additional markings on the forewing.

Male genitalia. Unknown.

Female genitalia. Anterior apophyses very short; membranous zone posterior from ostium bursae surrounded by two, strongly sclerotized plates; ductus bursae narrow, very long.

Early stages. Unknown.

Biology. Unknown except for the collecting dates (October or November).

Distribution. Cameroon.

Remarks. Only two females known so far. The placement in this subgenus is almost certain as the overall colouration and dark veins of the hindwing suggest close affinity to *R. invaria*.

181. *Rhipidarctia (Rhipidarctia) rhodosoma* KIRIAKOFF, 1957
(Pl. VI; ♂ Pl. 18)

Short diagnosis. Imago. Forewing length: 13 mm. Most similar to *R. flaviceps* and *R. aurora*. From the first it differs by the orange thorax and abdomen, from the second by the red suffusion in the interspace between forewing M_2 and A1-A2.

Male genitalia. Indistinguishable from those of the other species. See **Remarks** for *R. aurora*.

Female genitalia. Unknown.

Early stages. Unknown.

Biology. Unknown except for the collecting dates (April, August).

Distribution. DRC (Sankuru).

Remarks. Only one male is known. It is most possibly only a colour form of *R. aurora*.

182. *Rhipidarctia (Rhipidarctia) xenops* (KIRIAKOFF, 1957)
(Pl. VI; ♂ Pl. 18)

Short diagnosis. Imago. Forewing length: 12 mm. Body dark yellow-brown from above; wings dark ochraceous.

Forewing very narrow. Base and inner margin of forewing and entire hindwing blackish brown.

Male genitalia. As in subgenus *Rhipidarctia*; juxta prominent; uncus moderately sclerotized, tipped with small hook directed ventrally.

Female genitalia. Unknown.

Early stages. Unknown.

Biology. Unknown except for the collecting dates (April).

Distribution. Ghana.

Remarks. Only two males are known. Described by KIRIAKOFF in the new, monotypic genus *Takwa*.

Rhipidarctia INCERTAE SEDIS

183. *Rhipidarctia rubrovitta* (AURIVILLIUS, 1904)
(Pl. VI; ♀ Pl. 36)

Short diagnosis. Imago. Forewing length: 17-20 mm. Large species. Body olive with red blotches on forewing, red hindwing, and red narrow transverse band along distal edge of each tergite.

Male genitalia. Unknown.

Female genitalia. Anterior apophyses elongate, narrow, sclerotized; 7th and especially 8th sternite with setae less numerous than in other species of genus; ductus bursae elongate, narrow.

Early stages. Unknown.

Biology. Unknown except for the collecting date (May).

Distribution. Cameroon.

Remarks. The species has been wrongly attributed to *Balacra* by ROTHSHILD (1912b) and later by KIRIAKOFF (1960). Only three females are known until now.

184. *Rhipidarctia silacea* (PLÖTZ, 1880)

Short diagnosis. Imago. Forewing length: 21 mm.

Male genitalia. Unknown.

Female genitalia. Not examined

Early stages. Unknown.

Biology. Unknown except for the collecting date (March).

Distribution. Congo.

Remarks. The type was not located.

185. *Rhipidarctia syntomia* (PLÖTZ, 1880)

Short diagnosis. Imago. Forewing length: 14 mm.

Male genitalia. Unknown.

Female genitalia. Not examined.

Early stages. Unknown.

Biology. Unknown except for the collecting date (May).

Distribution. Ivory Coast.

Remarks. The type was not located.

Thyretes BOISDUVAL, 1847

186. *Thyretes buettikeri* WILTSHIRE, 1983
(Pl. VI; ♂ Pl. 18)

Short diagnosis. Imago. Forewing length: 12-13 mm. Externally indistinguishable from *T. negus*.

Male genitalia. According to WILTSHIRE (1983) resembling *T. negus* but differing in slightly convex costa of valva medially.

Female genitalia. Unknown.

Early stages. Unknown.

Biology. Unknown except for the collecting dates (September).

Distribution. Saudi Arabia, Yemen.

Remarks. This taxon is most probably conspecific with *T. negus* as the differences in the shape of the genitalia are very small. More material from the Arabian Peninsula is needed to check the range of morphological variation. So far only a few specimens have been collected. *Thyretes negus* has not yet been recorded from the Arabian Peninsula.

187. *Thyretes caffra* WALLENGREN, 1863
(Pl. VI; ♂ Pl. 18; ♀ Pl. 36)

Short diagnosis. Imago. Forewing length: 14-18 mm. Small, transparent, oval spot in cell always present, situated at base of veins CuA_1 and CuA_2; elongate blotch between anal vein and cell never extending beyond inner margin of next blotch.

Male genitalia. Indistinguishable from those of the other representatives of the genus; see **Remarks**.

Female genitalia. 7th sternite moderately sclerotized; ostium bursae not sclerotized; 8th sternite convex with deep, narrow, medial concavity, posteriorly from ostium bursae; see **Remarks**.

Early stages. Unknown.

Biology. Unknown except for the collecting dates (October – February).

Distribution. Mozambique, Namibia, RSA, Zambia.

Remarks. South African species. As in most genera the morphology of the male and female genitalia is very homogeneous, providing no characters for separating the species. Moreover, some of the species are known from only a few specimens, which makes it impossible to analyze the intraspecific variability in the genitalia. My examination of the vesica of four species did not bring any useful information. Fortunately, the species are easily separated (except *T. buettikeri*) in external characters such as the pattern and colouration of the wing and body.

188. *Thyretes cooremani* KIRIAKOFF, 1953
(Pl. VI)

Short diagnosis. Imago. Forewing length: 10 mm. The smallest of all *Thyretes* species. Indistinguishable from *T.*

signivenis except for the uniformly yellow head, thorax, and abdomen.

Male genitalia. Not examined.

Female genitalia. Unknown.

Early stages. Unknown.

Biology. Unknown except for the collecting dates (June).

Distribution. DRC.

Remarks. Only the holotype is known. It is most probably only a very small, unusually coloured specimen of *T. signivenis*. A preparation of the genitalia certainly would not resolve the problem of the identity of this taxon as the genitalia of all representatives of the genus are almost identical. The date of collecting is written on the label in Arabic numerals as follows: "7·[or -]6-1949". This suggest that the specimen was collected in June. In original description this information is interpreted by Kiriakoff as: "6-VII-49".

189. *Thyretes hippotes* (CRAMER, [1775-80])
(Pl. VI; ♂ Pl. 18; ♀ Pl. 36)

Short diagnosis. Imago. Forewing length: 18-24 mm. Together with *T. montana* it forms a pair of large, stout species restricted to Southern Africa. It can be separated from *T. montana* by the lack of yellow tinge on the wings and body. The colouration is restricted to white and blackish markings.

Male genitalia. Indistinguishable from those of the other representatives of the genus; see **Remarks** for *T. caffra*.

Female genitalia. 7th sternite strongly sclerotized; ostium bursae sclerotized; 8th sternite convex with prominent, deep, medial concavity, posterior from ostium bursae; see **Remarks** for *T. caffra*.

Early stages. Unknown.

Biology. The caterpillar feeds on *Pentzia incana* (karroo-bush) (Asteraceae). The pupation lasts 3-4 weeks and takes place in August and September (TAYLOR 1949). Adults collected in October and November.

Distribution. Angola.

Remarks. Type not located.

190. *Thyretes montana* BOISDUVAL, 1847
(Pl. VI; ♂ Pl. 18; ♀ Pl. 37)

Short diagnosis. Imago. Forewing length: 18-20 mm. Separated from *T. hippotes* by the yellowish colouration of the body.

Male genitalia. Indistinguishable from those of the other representatives of the genus; see **Remarks** for *T. caffra*.

Female genitalia. 8th sternite without concavity; indistinguishable from those of *T. negus*; see **Remarks** for *T. caffra*.

Early stages. Unknown.

Biology. Unknown except for the collecting dates (January, February, April, July).

Distribution. RSA.

Remarks. Type not located.

191. *Thyretes monteiroi* BUTLER, 1876
(Pl. VI; ♂ Pl. 18; ♀ Pl. 37)

Short diagnosis. Imago. Forewing length: 11-14 mm. One of the three small species with the head, thorax, and

103

abdomen mostly yellow. In contrast to *T. signivenis* and *T. cooremani*, the wings are dark with greatly reduced pale blotches; forewing blotch between veins M_1 and M_2 always absent.

Male genitalia. Indistinguishable from those of the other representatives of the genus; see **Remarks** for *T. caffra*.

Female genitalia. 7th sternite strongly sclerotized; ductus bursae elongate; see **Remarks** for *T. caffra*.

Early stages. Unknown.

Biology. Unknown except for the collecting dates (October-May).

Distribution. Angola, DRC.

192. *Thyretes negus* OBERTHÜR, 1878
(Pl. VI; ♂ Pl. 18; ♀ Pl. 37)

Short diagnosis. Imago. Forewing length: 10-20 mm. Middle-sized *Thyretes* with mostly greyish black body and extensive pale markings on fore wings. Most similar to *T. caffra* from which, except for the distribution, it can be separated by the usually larger semi-transparent blotch, located in upper corner of cell and elongate blotch between anal vein and cell always extending beyond inner margin of next blotch and in some populations reaching the outer margin of wing.

Male genitalia. Indistinguishable from the other representatives of the genus; see **Remarks** for *T. caffra*.

Female genitalia. As *T. negus*; see **Remarks** for *T. caffra*.

Early stages. BUTLER (1896) gives information on the egg cited from notes of the collector R. CRAWSHAY – "pale green ova". More detailed data are provided by TOWNSEND (1944) who gives short descriptions of the egg, caterpillar, and pupa. The caterpillar is densely covered with brown setae.

Biology. The food plants are probably different species of grasses (LE PELLEY 1959, SEVASTOPULO 1975, TOWNSEND 1944). The eggs are laid in batches on leaves of grass. Pupation takes place in a loose cocoon among dead grass. Adults collected throughout the year.

Distribution. DRC, Ethiopia, Kenya, Liberia, Malawi, Namibia, Tanzania, Togo, Zimbabwe.

Remarks. Extremely variable in size and colouration and very common in the whole of tropical Africa. The form (*T. buettikeri*) described from the Arabian Peninsula may belong to this species as well.

Female genitalia illustrated here are taken from one of two syntypes of *Thyretes misa* STRAND, 1911. Both of them are females. One of them bearing small, rectangular, red label with printed word "Type" is here selected and designated as LECTOTYPE. The second female is labeled as PARALECTOTYPE. Genitalia of this specimen are depicted on plate 37.

193. *Thyretes signivenis* HERING, 1937
(Pl. VI; ♂ Pl. 18)

Short diagnosis. Imago. Forewing length: 11-13 mm. Head, thorax, and abdomen yellow except for creamy patagia and tegulae; pale blotch between veins M_1 and M_2 of forewing always present.

Male genitalia. Indistinguishable from those of the other representatives of the genus; see **Remarks** for *T. caffra*.

Female genitalia. Unknown.

Early stages. Unknown.

Biology. Unknown except for the collecting dates (October-April).

Distribution. DRC.

Remarks. So far the female is unknown. Possibly only an individual colour form of *T. negus*.

Thyretini INCERTAE SEDIS

194. *Pseudmelisa demiavis* KAYE, 1919
(Pl. VI; ♀ Pl. 37)

Short diagnosis. Imago. Forewing length: 15 mm.

Male genitalia. Unknown.

Female genitalia. Ductus bursae short, broad, gradually narrowed towards corpus bursae.

Early stages. Unknown.

Biology. Unknown except for the collecting date of holotype (October or November).

Distribution. Cameroon.

Remarks. Only one female is known until now.

REFERENCES

Alibert, H. 1951. Les insectes vivant sur les cacaoyers en Afrique occidentale. *Mémoires de l'Institut Francais d'Afrique Noire* **15**: 1-174.

Aurivillius, C. 1881. Om en samling Fjärilar från Gaboon. *Entomologisk Tidskrift* **2**: 38-47.

Aurivillius, C. 1892. Verzeichniss einer vom Herrn Fritz Theorin aus Gabun und dem Gebiete des Camerunflusses heimgebrachten Schmetterlingssammlung, II. Heterocera. *Entomologisk Tidskrift* **13**: 181-200.

Aurivillius, C. 1898. Diagnosen neuer Lepidopteren aus Afrika, IV. *Entomologisk Tidskrift* **19**: I77-186.

Aurivillius, C. 1900. Verzeichniss einer von den Herren Missionären E. Laman und W. Sjöholm bei Mukinbungu am unteren Congo zusammengebrachten Schmetterlingssammlung. *Öfversigt af Kongl. Vetenskaps-Akademiens Förhandlingar* **9**: 1039-1058.

Aurivillius, C. 1904. Beiträge zur Kenntnis der Insektenfauna von Kamerun No 11. Lepidoptera Heterocera 2. *Arkiv för zoologi* **2**(4): 1-68.

Aurivillius, C. 1905a. Lieutnant A. Schulultzes Sammlung von Lepidopteren aus West-Afrika. *Arkiv för Zoologi* **2**(12): 1-47.

Aurivillius, C. 1905b. Verzeichnis von Lepidopteren, gesammelt bei Mukimbungu am unteren Kongo von Herrn E. Laman. Zweite und dritte Sendung. *Arkiv för Zoologi* **3**(1): 1-16.

Aurivillius, C. 1910. Schmetterlinge Gesammelt in Westafrika von Leonardo Fea in den Jahren 1897-1902. *Annali del Museo Civico di Storia Naturale di Genova* (3 ser.) **4**: 494-530.

Aurivillius, C. 1910b. Wissenschaftliche Ergebnisse der Schwedischen Zoologischen Expedition nach dem Kilimandjaro, dem Meru und den Umgebenden Massaisteppen Deutsch-Ostafrikas 1905-1906 unter leitung von Prof. dr. Yngve Sjöstedt. 2. Band; 9. Lepidoptera, 56 pp. Stockholm.

Aurivillius, C. 1921. Sammlungen der schwedischen Elgon-Expedition im Jahre 1920. 1. Lepidopteren. *Arkiv för Zoologi* **14**(5): 1-27.

Aurivillius, C. 1925a. Zoological result of the Swedish expedition to Central Africa 1921. Insecta 12. Lepidoptera 1. *Arkiv för Zoologi* **17a**(32): 1-20.

Aurivillius, C. 1925b. Lepidoptera IV. Teil. [in:] Schubotz H. *Ergebnisse der Zweiten Deutschen Zentral-Afrika Expedition 1910-1911 unter Führung Adolf Friedrichs, Herzogs zu Mecklenburg.*, Volume I, Lieferung 18: 1243-1359; Leipzig, Germany: Klinkhardt & Biermann.

Barns, T. A. 1923. Across the Great Craterland to the Congo. 276 pp. London. Ernest Benn Limited.

Barrett, F. 1902. Further Notes on South African Lepidoptera. *TheEntomologist's Monthly Magazine* (2)**13**: 138-143.

Bendib, A. & Minet, J. 1998. Female pheromone glands in Arctiidae (Lepidoptera). Evolution and phylogenetic significance. *Comptes Rendus de l'Academie des Sciences, Serie III - Sciences de la Vie* **321**: 1007-1014.

Berio, E. 1933. Note sui Lepidotteri. *Metarctia lateritia* H-S. e *Automolis unicolor* Obth. *Annali dell Museo Civico di Storia Naturale di Genova* **56**: 367-368.

Berio, E. 1935. Spedizione zoologica del Marchese Saverio Patrizi nel Basso Giuba e nell'Oltregiuba. Nuove specie di eteroceri. Amatidae, Arctiidae, Noctuidae. *Annali del Museo Civico di Storia Naturale di Genova* **58**: 56-65.

Berio, E. 1938. Spedizione zoologica del Marchese Saverio Patrizi nel Basso Giuba e nell'Oltregiuba. Lista dei Lepidotteri Eteroceri con note e diagnosi di eteroceri Africani. *Annali dell Museo Civico di Storia Naturale di Genova* **58**: 189-203.

Berio, E. 1939. Contributi alla conoscenza dei Lepidotteri Eteroceri dell'Eritrea I. Lista delle specie con descrizioni delle nuove entita raccolte negli anni 1934 al 1937 dal sig. Francesco Vaccaro. *Memorie della Societa Entomologica Italiana* **17**: 47-62.

Berio. E. 1940a. Lepidotteri Raccolti dal col. Mommeret ad Asmara nel Luglio-Ottobre 1934 con descrizione di una nuova Asticta (Noctuidae). *Bollettino della Societa Entomologica Italiana* **72**(3): 42-44.

Berio, E. 1940b. Contributo alla Conoscenza dei Lepidotteri Eteroceri dell'Eritrea. VI Eteroceri Raccolti dal cap.

Richini ad Adi-abuna (Adua) nel Marzo 1940. *Bollettino della Societa Entomologica Italiana* **72**(10): 161-165.

Berio, E. 1941. Elenco di Lepidotteri Eteroceri raccolti da Querci-Romei in Somalia con Diagnosi di nuove specie. *Memorie della Societa Entomologica Italiana* **20**: 118-124.

Berio, E. 1943. Contributi allo studio dei Lepidotteri Eteroceri dell'Eritrea. VII. Euchromiidae, Arctiidae, Agaristidae, Lymantriidae, Lasiocampidae, Noctuidae raccolte dal Sig. G. Vaccaro nel 1938. *Annali del Museo Civico di Storia Naturale Giacomo Doria* **61**: 176-190.

Bethune-Baker, G. T. 1911. Descriptions of new African Heterocera. *The Annals and Magazine of Natural History* (8)**7**: 530-552.

Bethune-Baker, G. T. 1927. Descriptions of new species of Heterocera from Africa and the East. *The Annals and Magazine of Natural History* (9) **20**: 321-334.

Boisduval, J. B. A. D. 1847. Catalogue des Lépidoptères recueillis par M. Delegorgue. [in:] Delegorgue, A. *Voyage dans l'Afrique australe 1838-1844*, **2**: 584-602, Paris.

Butler, A. G. 1876. Notes on the Lepidoptera of the Family Zygaenidae, with Descriptions of new Genera and Species. *Journal of the Linnean Society of London* **12**: 342-407.

Butler, A. G. 1877. Illustrations of Typical Species of Lepidoptera Heterocera in the Collection of the British Museum 1: i-xiii+1-62, London.

Butler, A. G. 1897. On two collections of Lepidoptera made by Mr. Crawshay in Nyasa-land, *Proceedings of the Zoological Society of London* **1896**: 817-850.

Condamin, M. & Roy, R. 1968. II. Vue d'ensemble sur la faune et le peuplement animal. Le Parc national du Niokolo-Koba, fasc. 3. *Mémoires de l'IFAN* **84**: 19-67.

Cramer, P. 1775-80. DeUitlandsche Kapellen voorkomende in de drie Waereld-deelen Asia, Africa en America. [Papilions exotiques des trois parties du Monde, l'Asie, l'Afrique et l'Amérique]. Vol. 3: 1-176, pls. CXCIII-CCLXXXVIII. S. J. Baalde, Amsterdam; B. Wild, Utrecht.

Debauche, A. 1938. Amatidae et Lithosiidae nouveaux ou peu connus. *Bulletin du Musée royal d'Histoire naturelle de Belgique* **14**(9): 1-21.

Debauche, A. 1942. Lépidoptères Hétérocères. *Exploration du Parc National Albert, Mission G. E. de Witte (1933-35)* **41**: 1-28.

De Freina, J. J. & Witt, T. T. 1987. Die Bombyces und Sphinges der Westpalaearktis. Edition Forschung & Wissenschaft Verlag GmbH, 708 pp. Munchen.

Delegorgue, A. 1847. Voyage dans l'Afrique Australe. **2**, 622 pp. Paris.

Druce, H. 1887. Descriptions of some new Species of Lepidoptera Heterocera mostly from tropical Africa. *Proceedings of the Zoological Society of London* **1887**: 668-686.

Druce, H. 1910. Descriptions of some new Species of Heterocera from tropical Africa. *The Annals and Magazine of Natural History* (8)**5**: 393-402.

Dufrane, A. 1936. Hétérocères. *Bulletin et Annales de la Société Entomologique de Belgique* **76**: 121-130.

Dufrane, A. 1940-44. Lépidoptères du Kivu (2e note). *Bulletin et Annales de la Société Entomologique de Belgique* **80**: 129-134.

Dufrane, A. 1945. Lépidoptères du Kivu (3e note). *Bulletin et Annales de la Société Entomologique de Belgique* **81**: 90-143.

Dufrane, A. 1952. Corrections. *Bulletin et Annales de la Société Entomologique de Belgique* **88**: 24.

Dyar, H. G. 1899. New Species of Syntomidae. *Journal of the New York Entomological Society* **7**: 174-176.

Fawcett, J. M. 1903. Notes on the Transformations of some South African Lepidoptera (continued from volume XV). *Transactions of the Zoological Society of London* **17**(2): 165-190.

Fawcett, J. M. 1915. Notes on a collection of Heterocera made by Mr. W. Feather in British East Africa, 1911-1912. *Proceedings of the Zoological Society of London* **1915**: 91-113.

Felder, C., Felder, R. & Rogenhofer, A. F. 1874. Reise der Österreichischen Fregatte Novara um die Erde (Zoologischer Theil) Band 2 (Abtheilung 2 Lepidoptera) Heft 4: 10 pp; pl. 99, Vienna.

Fontaine, M. 1992. Description des premiers états de quelques espèces de Lépidoptères Hétérocères Africains.

Lambillionea **92**(2): 113-115.

Gaede, M. 1926. Amatiden des Berliner Zoologischen Museums (Lep.). *Deutsche Entomologische Zeitschrift* **1926**(2): 113-136.

Gohrbandt, I. 1939. Ein neuer Typus des Tympanalorgans der Syntomiden. *Zoologischer Anzeiger* **126**(1-2): 107-116.

Grünberg, K. 1907. Einige neue afrikanische Heterocera. *Deutsche Entomologische Zeitschrift* **1907**(4): 431-437.

Hacker, H. H. 1999. Systematic List of Lepidoptera of the Arabian Paninsula with a survey of the spread with special reference to the fauna of Yemen. *Esperiana* **7**: 15-237.

Hampson, G. F. 1898. Catalogue of the Lepidoptera Phalaenae in the British Museum. **1**: XXII + 559 pp.

Hampson, G. F. 1901. New Species of Syntomidae and Arctiidae. *The Annals and Magazine of Natural History* (7)**8**: 165-186.

Hampson, G. F. 1902. The Moths of South Africa (Part I). *Annals of the South African Museum* **2**: 33- 66.

Hampson, G. F. 1905. Descriptions of new Genera and Species of Syntomidae, Arctiidae, Agaristidae and Noctuidae. *The Annals and Magazine of Natural History* (7)**15**: 425-452.

Hampson, G. F. 1907. Descriptions of new Genera and Species of Syntomidae, Arctiidae, Agaristidae and Noctuidae. *The Annals and Magazine of Natural History* (7)**19**: 221-257.

Hampson, G. F. 1909. Descriptions of new Genera and Species of Syntomidae, Arctiidae, Agarstidae and Noctuidae. *The Annals and Magazine of Natural History* (8)**4**: 344-388.

Hampson, G. F. 1910. Zoological collections from northern Rhodesia and adjacent territories: Lepidoptera Phalaenae. *Proceedings of the Zoological Society of London* **1910**: 388-510.

Hampson, G. F. 1911. Descriptions of new Genera and Species of Syntomidae, Arctiidae, Agaristidae and Noctuidae. *The Annals and Magazine of Natural History* (8)**8**: 393-445.

Hampson, G. F. 1914a. Catalogue of the Lepidoptera Phalaenae in the British Museum 1: (suppl.) XXVIII + 858 pp.

Hampson, G. F. 1914b. Two new species of wood-boring moths from West Africa. *Bulletin of the Entomological Research*. **5**: 245.

Hampson, G. F. 1916. [in:] Poulton, E. B. On a collection of moths made in Somaliland by Mr. W. Feather. *Proceedings of the Zoological Society of London* **1916**: 91-182.

Hampson, G. F. 1918. Descriptions of new Genera and Species of Amatidae, Lithosiidae and Noctuidae. *Novitates Zoologicae* **25**: 93-217.

Hampson, G. F. 1920. On new Genera and Species of Lepidoptera Phalaenae with the characters of two new families. *Novitates Zoologicae* **26**: 253-282.

Hering, M. 1932. Neue Heteroceren aus Afrika. *Revue de Zoologie et de Botanique Africaine* **22**(1): 102-117.

Hering, M. 1937. Eine neue Thyretes (Lep. Syntomid.) aus dem Congo-Gebiet, Thyretes signivenis (spec. nov.). *Revue de Zoologie et de Botanique Africaine* **29**(3): 229.

Herrich-Schäffer, G. A. W. 1850-1858. Sammlung neuer oder wenig bekannter aussereuropäischer Schmetterlinge, 84 pp., 120 pl., Regensburg.

Holland, W. J. 1892. Descriptions of some new Species of African Lepidoptera. *The Entomologist* (suppl.) **25**: 89- 95.

Holland, W. J. 1893. Descriptions of new Species and Genera of West African Lepidoptera. *Psyche* **6**: 373-376, 393-400, 411-418, 431-434, 451-454, 469-476, 487-490, 513-520, 531-538, 549-552, 565- 568.

Holland, W. J. 1896. List of the Lepidoptera collected in Eastern Africa by Dr. W. L. Abbott, with descriptions of some apparently new Species. *Proceedings of the United States National Museum* **18**: 229-258.

Holland, W. J. 1898. Descriptions of new West African Heterocera. *Entomological News* **9**: 11-13.

Holland, W. J. 1920. Lepidoptera of the Congo, being a systematic List of the Butterflies and Moths collected by the American Museum of Natural History Congo Expedition, together with Descriptions of some hitherto undescribed Species. *Bulletin of the American Museum of Natural History* **43**: 109-364.

Holloway J. D. 1988. The moths of Borneo: family Arctiidae, subfamilies Syntominae, Euchromiinae, Arctiinae; Noctuidae misplaced in Arctiidae (Camptoloma, Aganainae). Malayan Nature Society, Kuala Lumpur.

Honrath E. G. 1892. Neue Rhopalocera. *Berliner entomologische Zeitschrift* **36**(2): 429-440.

Hübner J. (1816-[1826]). Verzeichniss bekannter Schmetterlinge. 431 pp. + Anzeiger 72 pp. [signature 1, pp. [1]-16 (1816); sign. 2-11, pp. 17-176 [1819]; sign. 12-13, pp. 177-208 [1820]; sign. 14-16, pp. 209-256 [1821]; sign. 17-19, pp. 257-304 [1823]; sign. 20-27, pp. 305-431 [1825]; Anzeiger, numbered separately 1-72 [1826]; published privately, Augsburg.

Hulstaert, G. 1923. Hétérocères nouveaux du Congo belge. *Revue Zoologique Africaine* **11**: 406-411.

Jacobson, N. J. & Weller S. J. 2002. A Cladistic Study of the Arctiidae (Lepidoptera) by Using Characters of Immatures and Adults. *Thomas Say Publications in Entomology: Monographs.* Lahman, Maryland.

Janse, A. J. T. 1945. On the South African species of *Metarctia*, with the description of a new species. *Journal of the Entomological Society of Southern Africa* **8**: 91-98.

Joicey, J. J. & Talbot, G. 1921. New Lepidoptera collected by Mr. T. A. Barns.; V. -New Heterocera. *Bulletin of the Hill Museum* **1**(1): 158-166.

Joicey, J. J. & Talbot, G. 1924. New forms of African Lepidoptera. *Bulletin of the Hill Museum* **1**(3): 539-564.

Jordan, K. 1904. Some new moths. *Novitates Zoologicae* **11**: 441-447.

Jordan, K. 1907. New African Zygaenidae. *The Entomologist* **40**: 121-127.

Jordan, K. 1936. Two new African Syntomidae. *Novitates Zoologicae* **34**: 292-293.

Kaye, W. J. 1918. Descriptions from the Joicey collection of new species of Syntomidae, Nymphalidae, Hesperiidae, and two genera of Syntomidae. *The Annals and Magazine of Natural History* (9)**2**: 225-232.

Kaye, W. J. 1919. New species and genera of Nymphalidae, Syntomidae and Sphingidae in the Joicey Collection. *The Annals and Magazine of Natural History* (9)**4**: 84-94.

Kirby, W. F. 1892. A Synonymic Catalogue of Lepidoptera Heterocera (moths). xi + 951 pp. London.

Kiriakoff, S. G. 1948. Recherches sur les organes tympaniques des Lépidoptères en rapport avec la classification. *Bulletin et Annales de la Société Entomologique de Belgique* **84**: 231-276.

Kiriakoff, S. G. 1949. Over de Phylogenie van de Thyretidae fam. nov. (Lepidoptera). *Natuurwetenschappelijk Tijdschrrift* **31**: 3-10.

Kiriakoff, S. G. 1952a. Anapisa gen. nov. (Arctiidae, Lepidoptera). *Revue de Zoologie et de Botanique Africaine* **46**: 173-178.

Kiriakoff, S. G. 1952b. Thyretidae nouveaux du Congo Belge (Lepidoptera : Notodontoidea). *Revue de Zoologie et de Botanique Africaine* **46**: 369-406.

Kiriakoff, S. G. 1952c. La position systématique de Balacra paradoxa M. Hering (Lepid. Thyretidae). *Biologisch Jaarboek* **19**: 74-79.

Kiriakoff, S. G. 1952d. Un Thyretide nouveau de la Côte d'Ivoire. *Revue Francaise d'Entomologie* **19**(3): 173-175.

Kiriakoff, S. G. 1953a. Arctiidae nouveaux du Musée Royal du Congo Belge (Suite et fin.). *Lambillionea* **53**(11-12): 93-96.

Kiriakoff, S. G. 1953b. Les Thyretide du Musée Royal du Congo Belge (Lepidoptera Notodontoidea). *Annales du Musée Royal du Congo Belge* (8)**26**: 1-91.

Kiriakoff, S. G. 1954a. Contributions à l'étude des Lépidoptères Hétérocères (deuxième note). *Bulletin de l'Institut Royal des Sciences Naturelles de Belgique* **30**(29): 1-10.

Kiriakoff, S. G. 1954b. Lepidoptera Heterocera. *Exploration du Parc National de l'Upemba* 26: 1-69.

Kiriakoff, S. G. 1954c.Communications. *Elsita* nom. nov. (Lepidoptera, Thyretidae). *Bulletin et Annales de la Société Entomologique de Belgique* **90**(1-2): 29.

Kiriakoff, S. G. 1955a. Contributions à l'etude de la faune entomologique du Ruanda-Urundi (Mission P. Basilewsky 1953). XXVI. Lepidoptera Thyretidae, Arctiidae et Zygaenidae. *Annales du Musée Royal du Congo Belge* (8)**36**: 263-266.

Kiriakoff, S. G. 1955b. Die Thyretidae (Lepidoptera: Notodontoidea) aus der Zoologischen Staatssammlung München. *Mitteilungen der Münchener Entomologischen Gesellschaft* **44-45**(1954-55): 250-266.

Kiriakoff, S. G. 1955c. On the female genitalic structures of the Thyretidae (Lepidoptera). *Biologisch Jaarboek* **22**: 115-128.

Kiriakoff, S. G. 1956a. Lépidoptères Hétérocères africains nouveaux du Musée Royal du Congo Belge. *Lambillionea* **56**(3-4): 22-27.

Kiriakoff, S. G. 1956b. Lépidoptères Hétérocères africains nouveaux du Musée Royal du Congo Belge (Suite et fin). *Lambillionea* **56**(5-6): 38-42.

Kiriakoff, S. G. 1957a. Lépidoptères nouveaux du Congo belge. *Revue de Zoologie et de Botanique Africaines* **55**: 269-284.

Kiriakoff, S. G. 1957b. Notes sur les Thyretidae (Lepidoptera Notodontoidea). *Bulletin et Annales de la Societe Entomologique de Belgique* **93**(5-6): 121-160.

Kiriakoff, S. G. 1957c. New Thyretidae (Lepidoptera Notodontoidea). *Tijdschrrift voor Entomologie* **100**(1): 95-114.

Kiriakoff, S. G. 1958. Thyretidae and Notodontidae. *British Museum (Natural History) Ruwenzori Expedition 1952* **1**: 41-53.

Kiriakoff, S. G. 1959a. On the typical specimens of Thyretidae (Lepidoptera : Notodontoidea) in the Zoological Museum, Humboldt University, Berlin. *Entomologische Berichten* **19**: 186-190.

Kiriakoff, S. G. 1959b. Notes sur les Notodontoidea (Lepidoptera) du Congo Belge. *Lambillionea* **59**(3-4): 24-34.

Kiriakoff, S. G. 1960a. Lépidoptères Hétérocères (partim) récoltés par P. Lippens en Jordanie et en Arabie Séoudite. *Bulletin de l'Institut Royal des Sciences Naturelles de Belgique* **36**(35): 1-12.

Kiriakoff, S. G. 1960b. Lepidoptera Fam. Thyretidae [in:] Wystman, P. Genera Insectorum **214e**: 1-66.

Kiriakoff, S. G. 1961a. Thyretidae nouveaux (Lepidoptera: Notodontoidea). *Lambillionea* **61**(1-2): 7-12.

Kiriakoff, S. G. 1961b. Die Thyretidae (Lepidoptera: Notodontoidea) der Zoologischen Staatssamlung München. *Mitteilungen der Münchner Entomologischen Gesellschaft* **51**: 96-110.

Kiriakoff, S. G. 1963a. Lepidoptera Heterocera (Partim). *Exploration du Parc National Albert* (2 ser.)**16**: 73-124.

Kiriakoff, S. G. 1963b. Lepidoptera Heterocera (Anthroceridae,Arctiidae, Thyretidae, Notodontidae). La réserve naturelle intégrale du Mont Nimba, fasc. V. *Mémoires de l'Institut Francais d'Afrique Noire* **66**: 401-409.

Kiriakoff, S. G. 1963c. The tympanic structures of the Lepidoptera and the taxonomy of the order. *Journal of the Lepidopterists' Society* **17**(1): 1-6.

Kiriakoff, S. G. 1965. Les Lépidoptères Hétérocères africains de la collection Abel Dufrane. Deuxième note: Arctiidae. *Bulletin de l'Institut Royal des Sciences Naturelles de Belgique* **41**(21): 1-17.

Kiriakoff, S. G. 1972. *Microbergeria luctuosa* gen. nov. sp. nov. (Lepidoptera Thyretidae) avec quelques remarques sur la distribution d'espèces déjà décrites de Thyretidae. *Lambillionea* **71**(11-12): 102-105.

Kiriakoff, S. G. 1973a. Neue oder wenig bekannte Thyretidae (Lepidoptera: Notodontoidea). *Mitteilungen der Münchner Entomologischen Gesellschaft (e.V.)* **63**: 49-66.

Kiriakoff, S. G. 1973b. Notodontoidea (Lepidoptera) aus der Staatssammlung München. *Mitteilungen der Münchner Entomologischen Gesellschaft (e.V.)* **63**: 67-92.

Kiriakoff, S. G. 1978. Thyretidae (Lepidoptera: Notodontoidea) from Mount Kenya. Scientific Report of the Belgian Mt. Kenya Bio-Expedition 1975, n° 12. *Revue de Zoologie Africaine* **92**(2): 513-517.

Kiriakoff, S. G. 1979. Neue aethiopische Notodontoidea (Lepidoptera) aus der Zoologischen Staatssammlung München. *Spixiana* **2**(3): 215-251.

Kitching I. J. & Rawlins J. E. 1998. The Noctuoidea [in:] Kristensen N. P. [ed.] Lepidoptera, vol. 1. Handbuch der Zoologie. De Gruyter, Berlin.

Kopij, G. 2005. Lepidoptera fauna of Lesotho. *Acta zoologica cracoviensia* **49B**(1-2): 137-180.

Lavabre, E. M. 1961. Protection des cultures de caféier, cacaoyer et autres plantes pérennes tropicales. Institut Francais du Café et du Cacao, Paris.

Lafontaine, J. D. & Fibiger M. 2006. Revised higher classification of the Noctuoidea (Lepidoptera). *Canadian*

Entomologist **138**: 610-635.

Le Cerf, F. 1922. Extrait du Voyage de M. le Baron Maurice de Rothschild en Ethiopie et en Afrique Orientale Anglaise (1904-05), Lépidoptères Hétérocères. Résultats scientifiques, Animaux Articulés. Paris: 387-482.

Le Pelley, R. H. 1959. Agricultural Insects of East Africa. East Africa High Commision. Nairobi. Kenya, 307 pp.

Mabille, P. 1890. Voyage de M. Ch. Alluaud dans le territoire d'Assinie en juillet et août 1886. Lépidoptères, avec des notes sur quelques autres espèces d'Afrique. *Annales de la Société Entomologique de France* (6)**10**: 17-51.

Medler, J. T. 1980. Insects of Nigeria. Check List and Bibliography. *Memoirs of the American Entomological Institute* **30**: 1-919.

Minet, J. 1982. Elements sur la systématique des Notodontidae et nouvelles données concernant leur étude faunistique à Madagascar. *Bulletin de la Société entomologique de France* **87**: 354-370.

Minet, J. 1986. Ebauche d'une classification moderne de l'ordre des Lépidoptères. *Alexanor* **14**: 291-313.

Möschler, H. B. 1887, Beiträge zur Schmetterlings-Fauna der Goldküste. *Abhandlungen der Senckenbergischen naturforschenden Gesellschaft* **15**: 49-97.

Nonveiller, G. 1984. Catalogue commenté et illustré des Insectes du Cameroun d'interêt agricole (apparitions, répartition, importance). Beograd, 210 pp.

Oberthür, C. 1878. III.-Etude sur la faune des Lépidoptères de la côte orientale d'Afrique (Hétérocères). *Etudes d'Entomologie* 3: 30-36.

Oberthür, C. 1880. Spedizione Italiana nell' Africa Equatoriale. Resultati zoologici. I. Lepidotteri. *Annali del Museo Civico di Storia Naturale di Genova* 15: 129-186.

Oberthür, C. 1909. Descriptions de Lépidoptères africains. *Etudes de Lépidoptérologie Comparée* 3: 93-99.

Oberthür, C. 1911. Lépidoptères hétérocères nouveaux ou peu connus de l'Afrique tropicale. *Annales de la Société Entomologique de France* 79: 467-472.

Pagenstecher, 1909. Die geographische Verbreitung der Schmetterlinge. Verlag von Gustav Fischer in Jena. 451 pp.

Pinhey, E.C.G. 1975. Moths of Southern Africa. Tafelberg Publishers Ltd., Cape Town. 273 pp.

Plötz, C. 1880. Verzeichniss der vom Professor Dr. R. Buchholtz in West-Africa gesammelten Schmetterlinge. *Stettiner Entomologische Zeitung* **41**: 76-88.

Przybyłowicz, Ł. & Dall'Asta U. 2003. Redescription of the genus Melisoides Strand, 1912 (Lepidoptera: Arctiidae) with notes on its synonymy. *Acta zoologica cracoviensia* **46**(4): 347-353.

Przybyłowicz, Ł. 2005. A new Afrotropical species of Metamicroptera Hulstaert, 1923 with the first record of M. rotundata Hulst. from Zambia (Lepidoptera: Arctiidae). *Acta zoologica cracoviensia* **48B**(1-2): 139-144.

Przybyłowicz, Ł. & Kühne L. 2008. Subfamily Syntominae (Noctuoidea, Arctiidae) [in:] Kühne L. [ed.] Butterflies and moth diversity of the Kakamega Forest (Kenya). Brandenburgische Universitätsdruckerei, Potsdam, 145-156.

Rebel, H. 1914. Wissenschaftliche Ergebnisse der Expedition R. Grauer nach Zentralafrika, Dezember 1909 bis Februar 1911. Lepidoptera. *Annalen des k. k. naturhistorischen Hofmuseums Wien* **28**: 219-294.

Romieux, J. 1934a. Description de Lépidoptères nouveaux du Haut-Katanga (Congo Belge). *Mitteilungen der Schweizerischen Entomologischen Gesellschaft* **16**(3): 139-147.

Romieux, J. 1934b. Description de Lépidoptères nouveaux du Haut-Katanga (Congo Belge). *Bulletin de la Société lépidoptérologique de Genève* **7**(3): 105-113.

Romieux, J. 1946. Rectification: Balacra paradoxa Hering, non mihi. *Mitteilungen der Schweizerischen Entomologischen Gesellschaft* **20**(3): 267.

Rothschild, W. 1910. Descriptions of new Syntomidae. *Novitates Zoologicae* **17**: 429-445.

Rothschild, W. 1912a. A synonymic catalogue of the Syntomid Genus Balacra Walk., with descriptions of new species. *Novitates Zoologicae* **19**: 119-122.

Rothschild, W. 1912b. Some unfigured Syntomidae, Aegeriadae, and Arctianae. *Novitates Zoologicae* **19**: 375-377.

Rothschild, W. 1913. Descriptions of two new Colias and some African Syntomidae. *Novitates Zoologicae* **20**:

187-188.

Rougeot, P.C. 1977. Mission entomologiques en Ethiopie 1973-75, fasc. 1. *Memoires du Museum National d'Histoire naturelle* (Zool.) **105**: 1-150.

Schaus, W. & Clements, W. G. 1893. On a collection of Sierra Leone Lepidoptera. London, 46 pp.

Seitz, A. 1926. Syntomidae [in:] Grossschmetterlinge der Erde, vol. XIV.

Sevastopulo, D.G. 1975. A list of the food plants of East African Macrolepidoptera. Part 2 – Moths (Heteroptera). *The Bulletin of the Amateur Entomologists' Society.* **34**: 175-184.

Snellen, P.C.T. 1886a. Note I. Nouvelle espèce des Syntomides (Lepidoptera Heterocera). Automolis kelleni, spec. nov. *Notes from the Leyden Museum* **8**: 1-2.

Snellen, P. C. T. 1886b. Automolis kelleni Snell. *Tijdschrift voor Entomologie* **29**: 224-227.

Stoll, C. 1780-82. DeUitlandsche Kapellen voorkomende in de drie Waereld-deelen Asia, Africa en America. [Papilions exotique des trois parties de Monde l'Asie, l'Afrique et l'Amerique]. Vol. 4: 1-252, pls. CCCV-CCCC. S. J. Baalde, Amsterdam; B. Wild, Utrecht.

Strand, E. 1911. Beschreibungen afrikanischer Lepidopteren insbesondere Striphnopterygiden. *Annales de la Societe Entomologique de Belgique* **55**: 145-164.

Strand, E. 1912. Zoologische Ergebnisse der Expedition des Herrn G. Tessmann nach Sud-Kamerun und Spanisch-Guinea. Lepidoptera I. *Archiv für Naturgeschichte* **78**(A6): 139-197.

Strand, E. 1916. Neue und wenig bekannte Nebenformen von Syntomididen. *Archiv für Naturgeschichte* **82**(A2): 79-86.

Strand, E. 1918. Ueber einige Lepidopteren der Familien Lycaenidae, Hesperiidae, Syntomididae und Sphingidae aus Belgisch Kongo. *Internationalen Entomologischen Zeitschrift Guben* **12**: 1-9.

Strand, E. 1920. Kritische Bemerkungen und Berichtigungen zum Supplementband I des Hampson'schen „Catalogue of the Lepidoptera Phalaenae". *Deutsche Entomologische Zeitschrift „Iris"* **34**: 217-226.

Szent-Ivany, J. 1942. Ostafrikanische Heteroceren (Lepidopt.) von Baron Bornemissza und Kittenberger gesammelt, nebst Beschreibung von 3 neuen Lasiocampiden. *Annales Hist.-Nat. Musei Nationalis Hungarici* **35**: 63-68.

Talbot, G. 1929a. New forms of African Lepidoptera. *Bulletin of the Hill Museum* **3**: 72-77.

Talbot, G. 1929b. New forms with two new Genera of African Heterocera. *Bulletin of the Hill Museum* **3**: 125-132.

Talbot, G. 1932. New forms of African Lepidoptera. *Bulletin of the Hill Museum* **4**: 170-177.

Taylor, J. 1949. Notes on Lepidoptera in the eastern Cape Province (part 1). *The Journal of the Entomological Society of Southern Africa* **12**: 78-95.

Tjönneland, A. 1962. Light trap catches of two species of Lepidoptera: Meganaclia perpusilla Walk. (Syntomidae) and Rhanidophora cinctigutta Walk. (Noctuidae) at Jinja Uganda. *Contributions from the Faculty of Science University College of Addis Ababa.* Ser. C (Zoology) **2**: 1-7.

Townsend, A. L. H. 1944. Further notes on the early stages of Heterocera bred in the Nakuru District. *Journal of the East Africa Natural History Society* **18**(1-2): 15-31.

Turati, E. 1924. Spedizione Lepidotterologica in Cirenaica 1921-1922. *Atti della Societa Italiana di Scienze Naturali e del Museo Civico di Storia Naturale in Milano* **63**: 21-191.

Vari, L. & Kroon, D. M. 1986. Southern African Lepidoptera. A series of cross-referenced indices. Lepidopterists Society of Southern Africa and the Transvaal Museum. Pretoria. i-x, 1-198.

Vari, L., Kroon, D. M. & Krüger, M. 2002. Classification and Checklist of the Species of Lepidoptera recorded in Southern Africa. Simple Solutions, Australia (Pty) Ltd, Chatswood-Australia, i-xxi + 385 pp.

Viette, P. & Fletcher, D. S. 1968. The Types of Lepidoptera Heterocera Described by P. Mabille. *The Bulletin of the British Museum (Natural History).* **21**(8): 391-425

Walker, F. *1854.* List of the specimens of Lepidopterous Insects in the collection of the British Museum *1: 1-278,* London.

Walker, F. 1855a. *List of the specimens of Lepidopterous Insects in the collection of the British Museum* **3**: 583-776, London.

Walker, F. 1855b. *List of the specimens of Lepidopterous Insects in the collection of the British Museum* **4**: 777-976, London.

Walker, F. 1855c. *List of the specimens of Lepidopterous Insects in the collection of the British Museum* **5**: 977-1257, London.

Walker, F. 1856. *List of the specimens of Lepidopterous Insects in the collection of the British Museum* **7**: 1508-1787, London.

Walker, F. 1864. *List of the specimens of Lepidopterous Insects in the collection of the British Museum* **31**(Suppl.): 1-321, London.

Walker, F. 1869. [in:] Chapman, T. II.-On some Lepidopterous Insects from Congo. *Proceedings of the Natural History Society of Glasgow* **1**(2): 325-378.

Wallengren, H. D. J. 1860. Lepidopterologische Mittheilungen. *Wiener Entomologische Monatschrift* **4**(6): 161-176.

Wallengren, H. D. J. 1863. Lepidopterologische Mittheilungen. *Wiener Entomologische Monatschrift* **7**(5): 137-151.

Wallengren, H. D. J. 1865. Heterocer-Fjarilar, Samlade i Kafferlandet af J. A. Wahlberg. *Kongliga Svenska Vetenskaps-Akademiens Handlingar* **5**(4): 1-83.

Warnecke, G. 1934. Ein zweiter Beitrag zur Kenntnis der Macrolepidopteren-Fauna Südwest-Arabiens. II Nachtfalter. *Mitteilungen der Münchener Entomologischen Gesellschaft* **24**:61-65.

Watson, A., Fletcher, D. S. & Nye I. W. B. 1980. Noctuoidea: Arctiidae, Cocytiidae, Ctenuchidae, Dilobidae, Dioptidae, Lymantriidae, Notodontidae, Strepsimanidae, Thaumetopoeidae & Thyretidae [in:] Nye I.W.B. [ed.] *The generic names of moths of the world* **2**. Trustees of the British Museum (Natural History), London, i-xvi + 228 pp.

Weidner, H. 1974. Deie Entomologische Sammlungen des Zoologischen Instituts und Zoologischen Museums der Universität Hamburg XI. Teil Insecta VIII. *Mittelungen aus dem Hamburgischen Zoologischen Museum und Institut* **70**: 181-266.

Wichgraf, C. 1922. Neue afrikanische Lepidopteren aus der Ertlschen Sammlung (Fortsetzung). *Internationale Entomologische Zeitschrift Guben* **15**(22):172-173.

Wiltshire, E. 1980a. The Larger Moths of Dhofar and their Zoogeographic Composition. *Journal of Oman Studies, Special Report* **2**: 187-216.

Wiltshire, E. 1980b. Insects of Saudi Arabia. Fam. Cossidae, Limacodidae, Sesiidae, Lasiocampidae, Sphingidae, Notodontidae, Geometridae, Lymantriidae, Nolidae, Arctiidae, Agaristidae, Noctuidae, Ctenuchidae. *Fauna of Saudi Arabia* **2**: 179-240.

Wiltshire, E. 1982. Insects of Saudi Arabia. Lepidoptera: Fam. Cossidae, Zygaeniidae, Sesiidae, Lasiocampidae, Bombycidae, Sphingidae, Thaumatopoeidae, Thyretidae, Notodontidae, Geometridae, Lymantriidae, Noctuidae, Ctenuchidae (Part 2). *Fauna of Saudi Arabia* **4**: 271-332.

Wiltshire, E. 1983. Insects of Saudi Arabia. Lepidoptera: Fam. Cossidae, Sphingidae, Thyretidae, Geometridae, Lymantriidae, Arctiidae, Agaristidae, Noctuidae, Ctenuchidae (Part 3). *Fauna of Saudi Arabia* **5**: 293-331.

Wiltshire, E. 1990. An Illustrated, Annotated Catalogue of the Macro-Heterocera of Saudi Arabia. *Fauna of Saudi Arabia* **11**: 91-250.

Zerny, H. 1912a. Syntomidae. [in:] *Lepidopterorum Catalogus* **7**: 1-179, Berlin.

Zerny, H. 1912b. Neue Heteroceren aus der Sammlung des k. k. naturhistorischen Hofmuseums in Wien. *Deutsche Entomologische Zeitschrift „Iris"* **26**: 119-124.

Zerny, H. 1931. Beiträge zur Kenntnis der Syntomiden (Fortsetzung). *Deutsche Entomologische Zeitschrift „Iris"* **45**: 1-27.

Additional literature cited in the text:

De Prins, W. & De Prins, J. 2005. Gracillariidae (Lepidoptera) – *In*: World Catalogue of Insects 6: 1-502.

Scoble, M. J. (ed.) 1999. Geometrid Moths of the World. – A Catalogue (Lepidoptera: Geometridae). 2 Volumes. – CSIRO Publications, Melbourne, NHM Publications, London, Apollo Books, Stenstrup, pp. i-xxv, 1-1016.

PLATE I

1. *Apisa (Apisa) arabica* WARNECKE, 1934; Yemen, „San'a"; 20.08.1931; ♂; coll. BMNH.

2. *Apisa (Apisa) canescens* WALKER, 1855; lectotype; "South Africa"; ♂; coll. BMNH.

3. *Apisa (Dufraneella) fontainei* KIRIAKOFF, 1959; holotype; Rwanda, "Kisenyi"; 26.04.1957; ♂; coll. RMCA.

4. *Apisa (Dufraneella) grisescens* (DUFRANE, 1945); paratype; DRC, "Kamituga"; 18.12.1939"; ♂; coll. RMCA.

5. *Apisa (Dufraneella) hildae* KIRIAKOFF, 1961; holotype; Namibia, "Okahandja"; 1.04.1956; ♂; coll. ZSM.

6. *Apisa (Dufraneella) rendalli* ROTHSCHILD, 1910; lectotype; Malawi, "Zomba, Upp[er] Shire R[iver]"; 10-12.[18]95; ♂; coll. BMNH.

7. *Apisa (Dufraneella) subcanescens* ROTHSCHILD, 1910; lectotype; Senegal, "Casamance"; ♂; coll. BMNH.

8. *Apisa (Parapisa) cinereocostata* HOLLAND, 1893; holotype; Gabon, "Kangwe, Ogove Riv[er]"; ♂; coll. CMNH.

9. *Apisa (Parapisa) subargentea* JOICEY & TALBOT, 1921; holotype; Rwanda, "Lake Tshohoa"; 08.[19]19; ♀; coll. BMNH.

11. *Automolis bicolora* (WALKER, 1856); syntype; RSA, "Natal"; ♂; coll. BMNH.

12. *Automolis crassa* (FELDER, 1874); RSA, "Stettyn's Kloof, Worcester Distr."; 17.10.1966; ♂; coll. TMSA.

13. *Automolis incensa* (WALKER, 1864); holotype; [no locality]; ♂; coll. BMNH.

14. *Automolis meteus* (STOLL, 1780-82); RSA, "Ngome Forest, 27.49S 31.25E"; 20-24.01.1883; ♂; coll. ISEA.

14a. *Automolis meteus* (STOLL, 1780-82); [no locality]; ♀; coll. BMNH.

15. *Automolis pallida* (HAMPSON, 1901); syntype; Kenya, "Kikuyu, Nairobi"; 6.03.1900; ♂; coll. BMNH.

16. *Balacra (B.) belga* KIRIAKOFF, 1954; holotype; DRC, "Lupweji"; 09.1937; ♀; coll. KBIN.

17. *Balacra (B.) caeruleifascia* WALKER, 1856; holotype of *germana* ROTHSCHILD, 1912; "Sierra Leone"; ♂; coll. BMNH.

17a. *Balacra (B.) caeruleifascia* WALKER, holotype of *ehrmanni* HOLLAND, 1893; "Liberia"; ♀; coll. CMNH.

18. *Balacra (B.) guillemei* (OBERTHÜR, 1911); syntype of *erubescens* JOICEY & TALBOT, 1924; DRC, "Luvua River, 85 miles north of Lake Mweru"; 04.[19]22; ♂; coll. BMNH.

19. *Balacra (B.) nigripennis* (AURIVILLIUS, 1904); holotype "Centralafrika"; ♀; coll. NHRS.

20. *Balacra (B.) rattrayi* (ROTHSCHILD, 1910); syntype; Uganda, "Entebbe"; 11.[1]902; ♂; coll. BMNH.

20a. *Balacra (B.) rattrayi* (ROTHSCHILD, 1910); Burundi, Kitega; 10.04.1964; ♀; coll. RMCA

21. *Balacra (Callobalacra) alberici* DUFRANE, 1945; paratype; DRC, "Kamituga"; 1.12.[19]39; ♂; coll. RMCA.

22 *Balacra (Callobalacra) jaensis* BETHUNE-BAKER, 1927; holotype; Cameroon, "Bitje, Ja River"; 2000 ft; ♀; coll. BMNH.

23. *Balacra (Callobalacra) rubrostriata* (AURIVILLIUS, 1898); DRC, "Lulua, Kapanga"; 12.1933; ♀; coll. RMCA.

24. *Balacra (Compsochromia) compsa* (JORDAN, 1904); holotype; Angola, "Pungo Andongo"; ♂; coll. BMNH.

24a. *Balacra (Compsochromia) compsa* (JORDAN, 1904); holotype of *melaena* HAMPSON, 1905; Uganda, "Masaka"; 6.11.[19]02; ♀; coll. BMNH.

25. *Balacra (Compsochromia) diaphana* KIRIAKOFF, 1957; holotype; Uganda, "Kawanda"; 27.10.1941; ♂; coll. BMNH.

25a. *Balacra (Compsochromia) diaphana* KIRIAKOFF, 1957; paratype; Uganda, "Kawanda"; 27.10.1941; ♀; coll. BMNH.

26. *Balacra (Daphaenisca) affinis* (ROTSCHILD, 1910); DRC, "W Kivu, South of Walikali, South side middle Lova Valley"; coll. BMNH.

Plate I

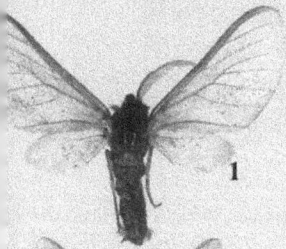

1

2

3

4

5

6

7

8

9

11

12

13

14

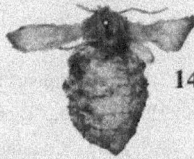

14a

15

16

17

17a

18

19

20

20a

21

22

23

24

24a

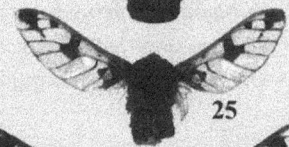

25

25a

26

PLATE II

27. *Balacra* (*Daphaenisca*) *daphaena* (Hampson, 1898); S Nigeria; ♂; coll. NHMV.

28. *Balacra* (*Heronina*) *herona* (Druce, 1887); Uganda, „Kampala"; ♂; coll. ISEA.

29. *Balacra* (*Lamprobalacra*) *elegans* Aurivillius, 1892; holotype; Cameroon; 10.03.[18]91; ♂; coll. NHRS.

30. *Balacra* (*Lamprobalacra*) *furva* Hampson, 1911; Ivory Coast, "Foret classee de la Bossematie, 20 km S ad Abengourou"; 09.1996; ♂; coll. ISEA.

31. *Balacra* (*Lamprobalacra*) *pulchra* Aurivillius, 1892; NW Cameroon, "Bascha"; ♂; coll. ZMHB.

32. *Balacra* (*Lamprobalacra*) *rubricincta* Holland, 1893; Cameroon, "Bipindihof"; 1914; ♂; coll. RNHL.

32a. *Balacra* (*Lamprobalacra*) *rubricincta* Holland, 1893; Zaire, "Coquilhatville"; 28.03.1925; ♀; coll. BMNH.

33. *Balacra* (*Pseudapiconoma*) *basilewskyi* Kiriakoff, 1953; holotype; DRC, "Bena-Dibele"; 1.11.1921; ♂; coll. RMCA.

34. *Balacra* (*Pseudapiconoma*) *batesi* (Druce, 1910); holotype; Cameroon, "Bitje, Ja River"; ♀; coll. BMNH.

35. *Balacra* (*Pseudapiconoma*) *flavimacula* Walker, 1856; holotype; Ghana, "Ashanti"; ♂; coll. BMNH.

35a. *Balacra* (*Pseudapiconoma*) *flavimacula* Walker, 1856; holotype of *B.* (*P.*) *testacea* (Aurivillius, 1881); Gabon; ♀; coll. NHRS.

36. *Balacra* (*Pseudapiconoma*) *fontainei* Kiriakoff, 1953; holotype; DRC, "Lusambo"; 28.09.1950; ♂; coll. RMCA.

37. *Balacra* (*Pseudapiconoma*) *haemalea* Holland, 1893; holotype; Gabon, "Kangwe, Ogove Riv[er]"; ♂; coll. CMNH.

38. *Balacra* (*Pseudapiconoma*) *humphreyi* Rothschild, 1912; holotype; Nigeria, "Ilesha"; ♂; coll. BMNH.

39. *Balacra* (*Pseudapiconoma*) *monotonia* (Strand, 1912); syntype; Equatorial Guinea, "Makomo, Ntumgebiet"; 14.04.[19]06; ♂; coll. ZMHB.

40. *Balacra* (*Pseudapiconoma*) *preussi* (Aurivillius, 1904); holotype; Cameroon, "Buea"; ♂; coll. NHRS.

40a. *Balacra* (*Pseudapiconoma*) *preussi* (Aurivillius, 1904); holotype of *speculigera* Grünberg, 1907; Cameroon, "Jaunde-Station"; ♀; coll. ZMHB.

41. *Bergeria bourgognei* Kiriakoff, 1952; holotype; Zaire; "Basoko"; 12.1936; ♂; coll. MRAC.

42. *Bergeria haematochrysia* Kiriakoff, 1952; holotype; Zaire, "Sankuru: Lusambo"; 20.08.1950; ♂; coll. MRAC.

42a. *Bergeria haematochrysia* Kiriakoff, 1952; paratype; Zaire, „Sankuru, Dimbelenge"; 12.03.1951; ♀; coll. MRAC.

43. *Bergeria octava* Kiriakoff, 1961; holotype; Zaire, "Ubangi, Region Bumba"; 11.09.1955; ♂; coll. MRAC.

44. *Bergeria ornata* Kiriakoff, 1959; holotype; Zaire, „Uele, Paulis"; 5.10.1957; ♂; coll. MRAC.

44a. *Bergeria ornata* Kiriakoff, 1959; Zaire, „Uele, Paulis"; 3.07.1956; ♀; coll. MRAC.

45. *Bergeria schoutedeni* Kiriakoff, 1952; holotype; Zaire, „Eala"; 04.1935; ♀; coll. MRAC.

46. *Bergeria tamsi* Kiriakoff, 1952; holotype; Zaire, "Equateur, Mondombe"; 20.08.1930; ♀; coll. MRAC.

47. *Cameroonia nigriceps* (Aurivillius, 1904); Cameroon, "Bitye, Ja River, 2000 ft"; 09-11.[19]11; ♂; coll. BMNH.

48. *Hippurarctia cinereoguttata* (Strand, 1912); holotype; Equatorial Guinea, "Nkolentangan"; ♂; coll. ZMHB.

49. *Hippurarctia ferrigera* (Druce, 1910); holotype of *vicina* Kiriakoff, 1953; Zaire, "Rwankwi"; 17.08.1947; ♂; coll. RMCA.

50. *Hippurarctia judith* Kiriakoff, 1959; holotype; DRC, "Uele, Paulis"; 26.03.1957; ♂; coll. RMCA.

51. *Hippurarctia taymansi* (Rothschild, 1910); holotype; DRC, "Kassai District, Congo F[ree]. St[ate]."; ♂; coll. BMNH.

Plate II

27

28

29

30

31

32

32a

33

34

35a

36

37

38

39

40

40a

41

42

42a

43

44

44a

45

46

47

48

49

50

51

PLATE III

52. *Lempkeella avellana* (Kiriakoff, 1957); holotype; DRC, "Eala"; 1935; ♂; coll. RMCA.

53. *Lempkeella dufranei* (Kiriakoff, 1952); holotype; DRC, "Elisabethville"; 1937; ♂; coll. RMCA.

54. *Lempkeella vanoyei* (Kiriakoff, 1952); holotype; DRC, "Bolombo"; 07.1938; ♂; coll. RMCA.

55. *Mecistorhabdia haematoessa* (Holland, 1893); holotype of *burgessi* Kiriakoff, 1957; Uganda, "Kigezi Dist., Impenetrable Forest, Kanungu"; ♂; coll. BMNH.

55a. *Mecistorhabdia haematoessa* (Holland, 1893); holotype; Gabon, "Kangwe, Ogove River"; ♀; coll. CMNH.

56. *Melisa croceipes* (Aurivillius, 1892); syntype of *atavistis* Hampson, 1911; Ghana, "Bibianaha"; ♂; coll. BMNH.

57. *Melisa diptera* (Walker, 1854); Ivory Coast, Lamto; 14.10.1966; ♂; coll. RMCA.

58. *Melisa hancocki* Jordan, 1936; syntype; Uganda, "Mabira Forest, nr. Jinja"; ♂; coll. OUM.

59. *Melisoides lobata* Strand, 1912; holotype of *collartorum* Kiriakoff, 1953; DRC, "Bokuma"; 28.12.1941; ♂; coll. RMCA.

59a. *Melisoides lobata* Strand, 1912; paratype of *collartorum* Kiriakoff, 1953; DRC, "Kapanga"; 03.1933; ♀; coll. RMCA.

60. *Metamicroptera christophi* Przybyłowicz, 2005; holotype; Zambia, „Nkana, N. Rhod." 11.1933; ♂; coll. TMSA.

61. *Metamicroptera rotundata* Hulstaert, 1923; lectotype of *paradoxa* Romieux, 1934; DRC, „Ht Katanga, Tshinkolobwe"; 10.11.[19]30; ♂; coll. MHNG.

63. *Metarctia (Collocalia) collocalia* Kiriakoff, 1957; holotype; Zimbabwe, "Vumba"; 7.11.1936.; ♂; coll. BMNH.

64. *Metarctia (Collocalia) debauchei* Kiriakoff, 1953; holotype; Burundi, "Usumbura"; ♂; coll. RMCA.

65. *Metarctia (Collocalia) dracoena* Kiriakoff, 1953; holotype; DRC, "Sankuru, Lusambo"; 2.05.1950; ♂; coll. RMCA.

66. *Metarctia (Collocalia) fuliginosa* Kiriakoff, 1953; holotype; DRC, "Lusambo"; 1.04.1950; ♂; coll. RMCA.

67. *Metarctia (Collocalia) jansei* Kiriakoff, 1957; holotype; RSA, "Natal"; ♂; coll. [BMNH].

68. *Metarctia (Collocalia) olbrechtsi* Kiriakoff, 1953; holotype; DRC, "Kat[anga]: Lubudi"; ♂; coll. RMCA.

69. *Metarctia (Collocalia) pavlitzkae* Kiriakoff, 1961; holotype; Tanzania, "Usambara Berge, Sakarani, 1500 m"; 15.11.1952; ♂; coll. ZSM.

70. *Metarctia (Collocalia) seydeliana* Kiriakoff, 1953; holotype; DRC, "Elisabethville"; 21.02.1934; ♂; coll. RMCA.

71. *Metarctia (Collocalia) tenebrosa* Le Cerf, 1922; holotype of *margaretha* Kiriakoff, 1957; Kenya, "Nairoba"; ♂; 04.[19]20; coll. BMNH.

72. *Metarctia (Hebena) cinnamomea* (Wallengren, 1860); RSA, "Ukhahlamba-Drakensberg Park, Didima Camp"; 28°56S 29°14E; 1400 m; 29.11.2004; ♂; coll. ISEA.

73. *Metarctia (Hebena) henrardi* Kiriakoff, 1953; holotype; DRC, "Terr. Kanda-Kanda: Gandajika"; 22.03.1948; ♂; coll. RMCA.

74. *Metarctia (Hebena) lateritia* (Herrich-Schaffer, 1850-1858); Tanzania, "Usa River, 3900 ft"; 09.1965-02.1966; ♂; coll. ISEA.

75. *Metarctia (Hebena) rubra* Walker, 1856; syntype of *kelleni* Snellen, 1886; Angola, "Z. Afrika, Humpata"; 17.02.[18]85; ♂; coll. RNHL.

76. *Metarctia (Hebena) subincarnata* Kiriakoff, 1954; paratype; DRC, "Lupweji"; 10.1937; ♂; coll. KBIN.

77. *Metarctia (Metarctia) alticola* Aurivillius, 1925; holotype of *rhodites* Kiriakoff, 1957; Rwanda, "Rugege Forest, Ruanda dist., Lake Kivu, 8000 ft"; 12.1921; ♂; coll. BMNH.

78. *Metarctia (Metarctia) atrivenata* Kiriakoff, 1956; holotype; Tanzania, "Mbeya T. T."; 03.1950; ♂; coll. RMCA.

79. *Metarctia (Metarctia) benitensis* Holland, 1893; syntype; Equatorial Guinea, "W. Africa, Benita"; ♂; coll. CMNH.

80. *Metarctia (Metarctia) brunneipennis* Hering, 1932; holotype; DRC, "Elisabethville"; 10.1926; ♀; coll. RMCA.

Plate III

52

60

71

53

61

7

54

55

63

73

55a

64

74

56

65

57

66

76

58

67

77

59

68

78

59a

69

79

70

80

PLATE IV

81. *Metarctia* **(***Metarctia***)** *burra* (SCHAUS in SCHAUS & CLEMENTS, 1893); holotype; [Sierra Leone]; ♂; coll. AMNH.

82. *Metarctia* **(***Metarctia***)** *burungae* DEBAUCHE, 1942; holotype; DRC, "Burunga, W. Kamatembe"; 14.03.1934; ♂; coll. RMCA.

83. *Metarctia* **(***Metarctia***)** *carmel* KIRIAKOFF, 1957; holotype; Ethiopia, "S.W. Abyssinia, Kambatta"; 30.04.[19]25; ♂; coll. BMNH.

84. *Metarctia* **(***Metarctia***)** *diversa* BETHUNE-BAKER, 1911; holotype; Angola, "N'Dalla Tando, 2700 ft"; 20.10.1908; ♂; coll. BMNH.

85. *Metarctia* **(***Metarctia***)** *fario* KIRIAKOFF, 1957; holotype; DRC, "Katenge, R. Katay"; 22.10.[19]24; ♂; coll. BMNH.

86. *Metarctia* **(***Metarctia***)** *flaviciliata* HAMPSON, 1907; lectotype; DRC, "Beni, Semliki, Ruwanzori"; 25.07.1906; ♂; coll. BMNH.

87. *Metarctia* **(***Metarctia***)** *flavicincta* AURIVILLIUS, 1900; holotype; DRC, "Congo"; ♂; coll. NHRS.

88. *Metarctia* **(***Metarctia***)** *flavivena* HAMPSON, 1901; syntype; Kenya, "Machakos"; 15.04.[18]98; ♂; coll. BMNH.

89. *Metarctia* **(***Metarctia***)** *flora* KIRIAKOFF, 1957; holotype; Rwanda, "Rugege Forest, Ruanda Dist., Lake Kivu, 7000 ft"; 12.1921; ♂; coll. BMNH.

90. *Metarctia* **(***Metarctia***)** *fontainei* KIRIAKOFF, 1953; holotype; DRC, "Sankuru, Katako-Kombe"; 17.12[?].1951; ♂; coll. RMCA.

91. *Metarctia* **(***Metarctia***)** *forsteri* KIRIAKOFF, 1955; holotype, "Kamerun"; ♂; coll. ZSM.

92. *Metarctia* **(***Metarctia***)** *fulvia* HAMPSON, 1901; lectotype; Kenya, "Athi-ya-Mawe, B.E. Africa"; 16.04. [18]99; ♂; coll. BMNH.

93. *Metarctia* **(***Metarctia***)** *fusca* HAMPSON, 1901; holotype; Kenya, "Kikuyu, Romoro, B.E. Africa"; 29.12.1900; ♂; coll. BMNH.

94. *Metarctia* **(***Metarctia***)** *galla* ROUGEOT, 1977; holotype; Ethiopia, "Reserve de Bale"; 3-5.11.1973; ♂; coll. MNHN.

95. *Metarctia* **(***Metarctia***)** *haematricha* HAMPSON, 1905; holotype; Ethiopia, "Abyssinia, Kutai Mecha"; 06.1902; ♂; coll. BMNH.

97. *Metarctia* **(***Metarctia***)** *hulstaertiana* KIRIAKOFF, 1953; holotype; DRC, "Equateur: Bokote"; 1927; ♂; coll. RMCA.

98. *Metarctia* **(***Metarctia***)** *inconspicua* HOLLAND, 1892; holotype; Tanzania, "Zanzibar"; ♂; coll. USNM.

99. *Metarctia* **(***Metarctia***)** *johanna* (KIRIAKOFF, 1979); paratype; Nigeria, "Ogoja"; 29.05.1974; ♂; coll. ZSM.

100. *Metarctia* **(***Metarctia***)** *kumasina* STRAND, 1920; lectotype; Ethiopia, "Abyssinia, Zegi Tsana"; 05-06.1902; ♂; coll. BMNH.

101. *Metarctia* **(***Metarctia***)** *lindemannae* KIRIAKOFF, 1961; paratype; Tanzania, "Tanganjika, Usambara-Berge, Sakarani, 1500 m"; 16.11.1952; ♂; coll. ZSM.

102. *Metarctia* **(***Metarctia***)** *longipalpis* HULSTAERT, 1923; holotype, DRC, "Elisabethville"; 11.1911; ♂; coll. RMCA.

103. *Metarctia* **(***Metarctia***)** *lugubris* GAEDE, 1926; holotype; Tanzania, "Ukinga-Berge"; 05.[18]99; ♂; coll. ZMHB.

104. *Metarctia* **(***Metarctia***)** *maria* KIRIAKOFF, 1957; holotype; Guinea, "Boukouni, N[ea]r. Macenta, 1750 ft"; 11.05.[19]26; ♂; coll. BMNH.

105. *Metarctia* **(***Metarctia***)** *metaleuca* HAMPSON, 1914; holotype; Liberia, „Nanna Kru"; 31.12.[19]10; ♂; coll. BMNH.

106. *Metarctia* **(***Metarctia***)** *morag* KIRIAKOFF, 1957; holotype; DRC, „Upper Oso River, NW Kivu, 4000 ft"; 02.[19]24; ♂; coll. BMNH.

107. *Metarctia* **(***Metarctia***)** *negusi* KIRIAKOFF, 1957; holotype; Ethiopia, „Abyssinie"; 1881; ♂; coll. BMNH.

108. *Metarctia* **(***Metarctia***)** *nigritarsis* BERIO, 1943; holotype; Eritrea, "Dorfu"; 20.10.1938; ♂; coll. MCSNG.

109. *Metarctia* **(***Metarctia***)** *noctis* DRUCE, 1910; syntype; Ethiopia, "Dire Daoua"; ♂; coll. BMNH.

110. *Metarctia* **(***Metarctia***)** *pallens* BETHUNE-BAKER, 1911; holotype; Angola, "N'Dalla Tando, 2700 ft"; 15.11.1908; ♂; coll. BMNH.

111. *Metarctia* **(***Metarctia***)** *paremphares* HOLLAND, 1893; syntype; Gabon, "Kangwe, Ogove Riv."; ♂; coll. CMNH.

112. *Metarctia* **(***Metarctia***)** *paulis* KIRIAKOFF, 1961; holotype; DRC, "Uele: Paulis"; 10.07.1958; ♂; coll. RMCA.

113. *Metarctia* **(***Metarctia***)** *phaeoptera* HAMPSON, 1909; holotype; DRC, "Upper Congo" 1907; ♀; coll. BMNH.

114. *Metarctia* **(***Metarctia***)** *priscilla* KIRIAKOFF, 1957; holotype; Ghana, "Bibianaha, 70 miles N.W. of Dimkwa"; 700 ft"; ♂; coll. BMNH.

115. *Metarctia* **(***Metarctia***)** *pulverea* HAMPSON, 1907; holotype, Uganda, "Ruwenzori, 6000 ft"; 30.01.[19]06; ♀; coll. BMNH.

Plate IV

81

82

83

84

85

86

87

88

89

90

91

92

93

94

95

97

98

99

100

101

102

103

104

105

106

107

108

109

110

111

112

113

114

115

PLATE V

116. *Metarctia (Metarctia) pumila* HAMPSON, 1909; lectotype; Sudan, "White Nile, Gondokoro"; ♂; coll. BMNH.

118. *Metarctia (Metarctia) rufescens* WALKER, 1855; syntype; RSA, "Natal"; ♂; coll. BMNH.

119. *Metarctia (Metarctia) saalfeldi* KIRIAKOFF, 1960 holotype [Ethiopia] "Villagio"; 06.[19]39; ♂; coll. ZSM.

120. *Metarctia (Metarctia) salmonea* KIRIAKOFF, 1957; holotype; Angola, "Dondo"; 21.02.[18]75; ♂; coll. BMNH.

121. *Metarctia (Metarctia) sarcosoma* HAMPSON, 1901; holotype; Kenya, "Machakos"; 15.04. [18]98; ♂; coll. BMNH.

122. *Metarctia (Metarctia) sheljuzhkoi* KIRIAKOFF, 1961; holotype; Ivory Coast, "Abidjan"; 4.09.1952; ♂; coll. ZSM.

123. *Metarctia (Metarctia) subpallens* KIRIAKOFF, 1956; holotype; Kenya, "Makueni"; 4.04.1954; ♂; coll. RMCA.

126. *Metarctia (Metarctia) tricolorana* WICHGRAF, 1922; holotype; Uganda, „Gulu"; ♂; coll. BMNH.

127. *Metarctia (Metarctia) unicolor* (OBERTHÜR, 1880); holotype; Ethiopia, „Presa in viaggio venando da Fin-Finni paese dei Galla"; 06.1878; ♂; coll. MCSNG.

128. *Metarctia (Metarctia) uniformis* BETHUNE-BAKER, 1911; holotype; Angola, "Malange"; 10.12.1909; ♂; coll. BMNH.

129. *Metarctia (Metarctia) upembae* KIRIAKOFF, 1954; holotype; DRC, „P.N.U., Kankunda, r. dr Lupiala, 1300 m"; 13-27.11.1947; ♂; coll. RMCA.

130. *Metarctia (Metarctia) venustissima* KIRIAKOFF, 1961; holotype; DRC, „Katanga: Kolwezi"; 18.11.1954; ♂; coll. RMCA.

131. *Metarctia (Metarctia) virgata* JOICEY & TALBOT, 1921; syntype; DRC, „Mikeno Mt., N. Kivu, 2400 ft"; 09.[19]19; ♂; coll. BMNH.

132. *Metarctia (Metarhodia) confederationis* KIRIAKOFF, 1961; holotype; RSA, "Natal, Karkloof"; 12.02.1929; ♂; coll. ZSM.

133. *Metarctia (Metarhodia) epimela* (KIRIAKOFF, 1979); holotype; Tanzania, "Mt. Meru, Momella, 1600-1800 m"; 1-10.02. [19]64; ♂; coll. ZSM.

134. *Metarctia (Metarhodia) heinrichi* KIRIAKOFF, 1961; holotype [Angola] "Prov. Nordcuanza, Canzele, 30 km nördl. Quiculungo"; 18.10.1957; ♂; coll. ZSM.

135. *Metarctia (Metarhodia) heringi* KIRIAKOFF, 1957; holotype; DRC, "Elisabethville"; 28.08.1948; ♂; coll. RMCA.

136. *Metarctia (Metarhodia) hypomela* KIRIAKOFF, 1956; holotype, Kenya, "Kakamega"; 06.1950; ♂; coll. RMCA.

137. *Metarctia (Metarhodia) insignis* KIRIAKOFF, 1959; holotype; Rwanda, "Kisenyi"; 17.04.1957; ♂; coll. RMCA.

138. *Metarctia (Metarhodia) jordani* KIRIAKOFF, 1957; holotype; Angola, "Mt. Moco, Luimbale, 1800-1900 m"; 16.03.1934; ♂; coll. BMNH.

139. *Metarctia (Metarhodia) nigricornis* DEBAUCHE, 1942; holotype; DRC, "Reg. Nyarusambo"; 27-30.06.1935; ♂; coll. RMCA.

140. *Metarctia (Metarhodia) rubribasa* BETHUNE-BAKER, 1911; holotype; Angola, "N'Dalla Tando, 2700 ft"; 22.12.1908; ♀; coll. BMNH.

141. *Metarctia (Metarhodia) rubripuncta* HAMPSON, 1898; holotype; Gabon, "Gaboon"; ♀; coll. BMNH.

143. *Metarctia (Pinheyata) quinta* KIRIAKOFF, 1973; Tanzania; „Mts Uluguru, Morogoro Campus Fac Agric"; 05-06.[19]71; ♂; coll. RMCA.

145. *Metarctia (Thyretarctia) didyma* KIRIAKOFF, 1957; holotype; Ghana, „Kumasi"; ♂; coll. BMNH.

146. *Metarctia (Thyretarctia) haematica* HOLLAND, 1893; holotype; Gabon, "Gaboon"; ♂; coll. CMNH.

146a. *Metarctia (Thyretarctia) haematica* HOLLAND, 1893; holotype of *M.(T.) haematosphages* HOLLAND, 1893; Gabon, "Kangwe, Ogove Riv."; ♀; coll. CMNH.

147. *Metarctia (Thyretarctia) infausta* KIRIAKOFF, 1957; holotype; DRC, "Kibali-Ituri: Nioka"; 23.03.1952; ♂; coll. RMCA.

148. *Metarctia (Thyretarctia) morosa* KIRIAKOFF, 1957; holotype; DRC, "Elisabethville"; 03.1912; ♂; coll. RMCA.

149. *Metarctia (Thyretarctia) schoutedeni* KIRIAKOFF, 1953; holotype; DRC, "Kivu: Burunga"; 12.1925; ♂; coll. RMCA.

151. *Neophemula vitrina* (OBERTHÜR, 1909); holotype of ssp. *angolensis* KIRIAKOFF, 1957; Angola, "Quicolungo, 120 km N of Lucala"; 04.1936; ♂; coll. BMNH.

152. *Owambarctia owamboensis* KIRIAKOFF, 1957; holotype; Namibia, "Tsumeb, region de l' Owambo"; ♂; coll. RMCA.

Plate V

116

129

139

118

140

119

130

141

120

131

143

121

132

145

133

122

146

134

123

146a

135

126

147

136

148

127

137

149

151

128

138

PLATE VI

153. *Owambarctia unipuncta* KIRIAKOFF, 1973; holotype; Tanzania, "Uruguru-Berge"; 14.12.1961; ♂; coll. ZSM.

154. *Paramelisa dollmani* HAMPSON, 1920; lectotype; Zambia, "Solwezi"; 12.09.1917; ♂; coll. BMNH.

154a. *Paramelisa dollmani* HAMPSON, 1920; paralectotype; Zambia, "Solwezi"; 27.11.1917; ♀; coll. BMNH.

155. *Paramelisa leroyi* KIRIAKOFF, 1953; holotype; DRC, "Rwankwi"; 25.02.1951; ♂; coll. RMCA.

156. *Paramelisa lophura* AURIVILLIUS, 1905; holotype; DRC, "Mukimbungu"; ♂; coll. NHRS.

157. *Paramelisa lophuroides* OBERTHÜR, 1911; holotype; Cameroon, "Johann Albrechts Höhe Station"; ♂; coll. BMNH.

158. *Pseudmelisa chalybsa* HAMPSON, 1910; holotype; DRC, "150-200 miles W. of Kambove"; 30.10.1907; ♀; coll. BMNH.

159. *Pseudmelisa rubrosignata* KIRIAKOFF, 1957; holotype; Malawi, "Nyasaland, Mlanje, Luchenya R[iver]"; 26.02.1914; ♀; coll. BMNH.

160. *Pseudothyretes carnea* (HAMPSON, 1898); holotype; "Angola"; ♀; coll. BMNH.

161. *Pseudothyretes erubescens* (HAMPSON 1901); holotype; Uganda, "Mile 478 on Uganda R[ai]l[wa]y "; 21.11.1900; ♂; coll. BMNH.

162. *Pseudothyretes kamitugensis* (DUFRANE, 1945); DRC, "N. Lac Kivu, Rwankwi"; 2.09.1948; ♂; coll. RMCA.

164. *Pseudothyretes nigrita* (KIRIAKOFF, 1961); Uganda, "Kallinzu Forest"; 7-10.01.1965; ♂; coll. ZSM.

165. *Pseudothyretes perpusilla* (WALKER, 1856); holotype; "S. Leone"; ♂; coll. BMNH.

166. *Pseudothyretes rubicundula* (STRAND, 1912); holotype; Equatorial Guinea, "Makomo, Ntumgebiet"; 14.05.[19]06; ♂; coll. ZMHB.

167. *Rhabdomarctia rubrilineata* (BETHUNE-BAKER, 1911); holotype; Angola, "N'Dalla Tando"; 23.11.1908; ♂; coll. BMNH.

167a. *Rhabdomarctia rubrilineata* yellow form; holotype of *similis* KIRIAKOFF, 1953; DRC, "Eala"; 08.1936; ♂; coll. RMCA.

168. *Rhipidarctia (Elsitia) cinctella* (KIRIAKOFF, 1953); holotype; DRC, "Lusambo"; 8.10.1950; ♂; coll. RMCA.

169. *Rhipidarctia (Elsitia) forsteri* (KIRIAKOFF, 1953); holotype; DRC, "Kivu, Rwankwi"; 20.08.1947; ♂; coll. RMCA.

170. *Rhipidarctia (Elsitia) lutea* HOLLAND, 1893; holotype; Gabon, "Kangwe, Ogove Riv."; ♀; coll. USNM.

171. *Rhipidarctia (Elsitia) pareclecta* (HOLLAND, 1893); holotype; Gabon, "Kangwe, Ogove Riv."; ♂; coll. CMNH.

172. *Rhipidarctia (Elsitia) saturata* KIRIAKOFF, 1957; holotype; DRC, "Sankuru: Katako-Kombe"; 9.12.1951; ♂; coll. RMCA.

173. *Rhipidarctia (Elsitia) subminiata* KIRIAKOFF, 1959; holotype; DRC, "Uele: Paulis"; 21.12.1957; ♂; coll. RMCA.

174. *Rhipidarctia (Hemirhipidia) postrosea* (ROTHSCHILD, 1913); paratype; Nigeria "Near Lagos, 1 mile from Oni [Camp]"; ♀; coll. OUM.

175. *Rhipidarctia (Rhipidarctia) aurora* KIRIAKOFF, 1957; paratype; DRC, „Eala"; 04.1935; ♂; coll. RMCA.

176. *Rhipidarctia (Rhipidarctia) conradti* OBERTHUR, 1911; syntype; Cameroon, „Johann-Albrechts Höhe"; ♀; coll. BMNH.

177. *Rhipidarctia (Rhipidarctia) crameri* KIRIAKOFF, 1961; holotype; Uganda, „Masindi"; 4.10.1959; ♂; coll. ZSM.

178. *Rhipidarctia (Rhipidarctia) flaviceps* HAMPSON, 1898; holotype; "Cameroons"; ♂; coll. BMNH.

178. *Rhipidarctia (Rhipidarctia) flaviceps* (WALKER, 1856); holotype; "West Africa"; ♀; coll. BMNH.

180. *Rhipidarctia (Rhipidarctia) miniata* KIRIAKOFF, 1957; holotype; Cameroon, "Bitje, Ja river"; 10-11.1913; ♀; coll. BMNH.

181. *Rhipidarctia (Rhipidarctia) rhodosoma* KIRIAKOFF, 1957; holotype; DRC, "Sankuru, Katako-Kombe"; 17.04.1952; ♂; coll. RMCA.

182. *Rhipidarctia (Rhipidarctia) xenops* (KIRIAKOFF, 1957); Nigeria, "Ilobi"; ♂; 04.[19]55; coll. BMNH.

183. *Rhipidarctia rubrovitta* (AURIVILLIUS, 1904); syntype; ♀; coll. NHRS.

186. *Thyretes buettikeri* WILTSHIRE, 1983; holotype; Saudi Arabia, "Fayfa"; 23.09.1981; ♂; coll. NHMB.

187. *Thyretes caffra* WALLENGREN, 1863; holotype; RSA, "Caffraria"; ♂; coll. NHRS.

188. *Thyretes cooremani* KIRIAKOFF, 1953; holotype; DRC, "Leopoldville"; 7.06.1949; ♂; coll. RMCA.

189. *Thyretes hippotes* (CRAMER, [1775-80]); RSA, "Noordhoek, Cape of Good Hope"; 8.10.[..]61; ♂; coll. BMNH.

190. *Thyretes montana* BOISDUVAL, 1847; RSA, "Transvaal"; ♀; coll. NHMW.

191. *Thyretes monteiroi* BUTLER, 1876; holotype of *angolensis* GAEDE, 1926; Angola, "Quisoll, 23 km v. Malange"; ♂; coll. ZMHB.

192. *Thyretes negus* OBERTHÜR, 1878; holotype; Ethiopia, "Abyss."; ♂; coll. BMNH.

193. *Thyretes signivenis* HERING, 1937; holotype; DRC, "Elisabethville"; 28.03.1936; ♂; coll. RMCA.

194. *Pseudmelisa demiavis* KAYE, 1919; holotype; Cameroon, "Bitje, Ja River"; 10-11.1912; ♀; coll. BMNH.

Plate VI

153

154

154a

155

156

157

158

159

160

161

162

164

165

166

167

167a

168

169

170

171

172

173

174

175

176

177

178

179

180

181

182

183

186

187

188

1

1

191

19

193

19

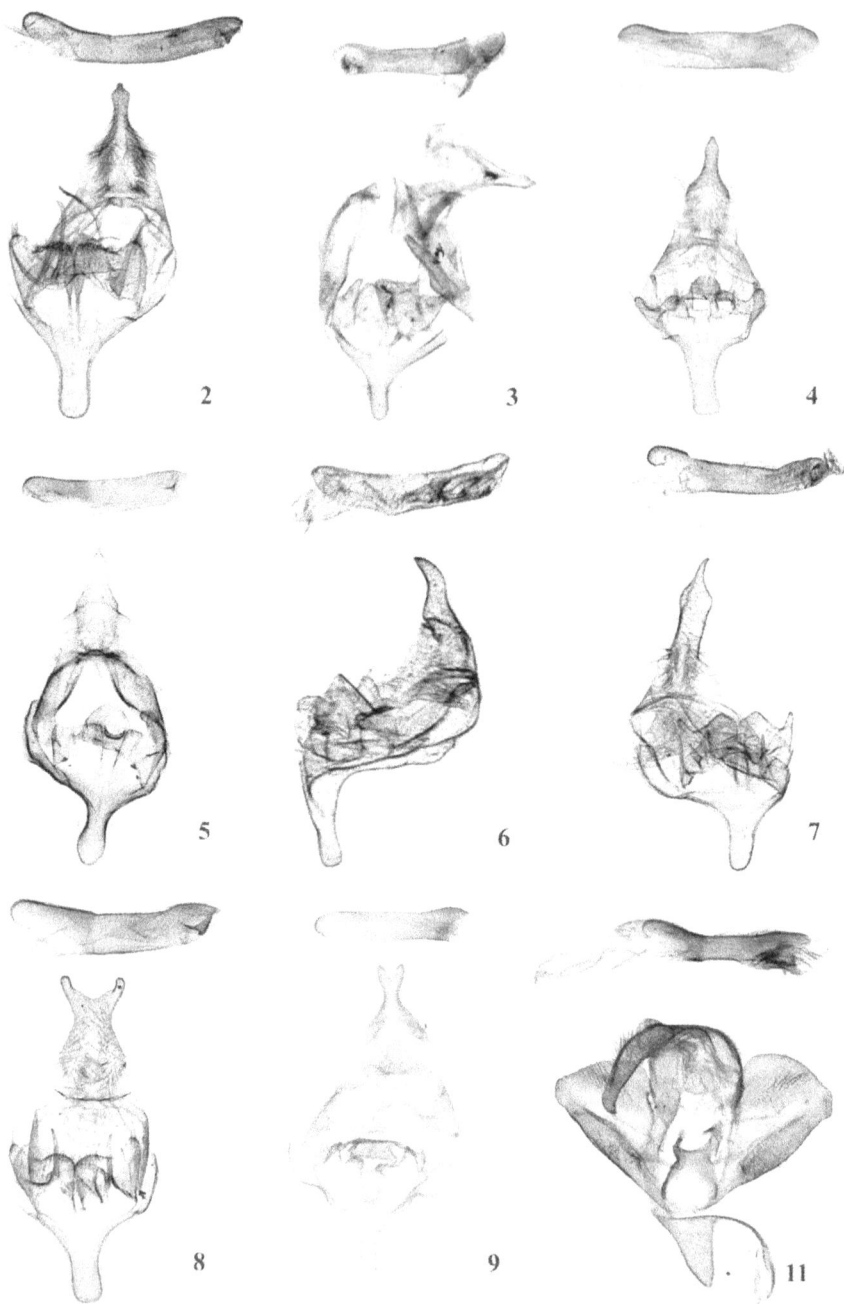

Plate 1

2. *Apisa (Apisa) canescens* WALKER, 1855; lectotype; coll. BMNH. - 3. *Apisa (Dufraneella) fontainei* KIRIAKOFF, 1959; holotype; coll. RMCA. - 4. *Apisa (Dufraneella) grisescens* (DUFRANE, 1945); holotype; coll. KBIN. - 5. *Apisa (Dufraneella) hildae* KIRIAKOFF, 1961; holotype; coll. ZSM. - 6. *Apisa (Dufraneella) rendalli* ROTHSCHILD, 1910; lectotype; coll. BMNH. - 7. *Apisa (Dufraneella) subcanescens* ROTHSCHILD, 1910; lectotype; coll. BMNH. - 8. *Apisa (Parapisa) cinereocostata* HOLLAND, 1893; holotype; coll. CMNH. - 9. *Apisa (Parapisa) subargentea* JOICEY & TALBOT, 1921; DRC, "Kibali-Ituri, Nioka"; 27.11.1953; coll. RMCA. - 11. *Automolis bicolora* (WALKER, 1856); RSA, "Balgowan, Natal"; 20.01.1950; coll. ZSM.

126

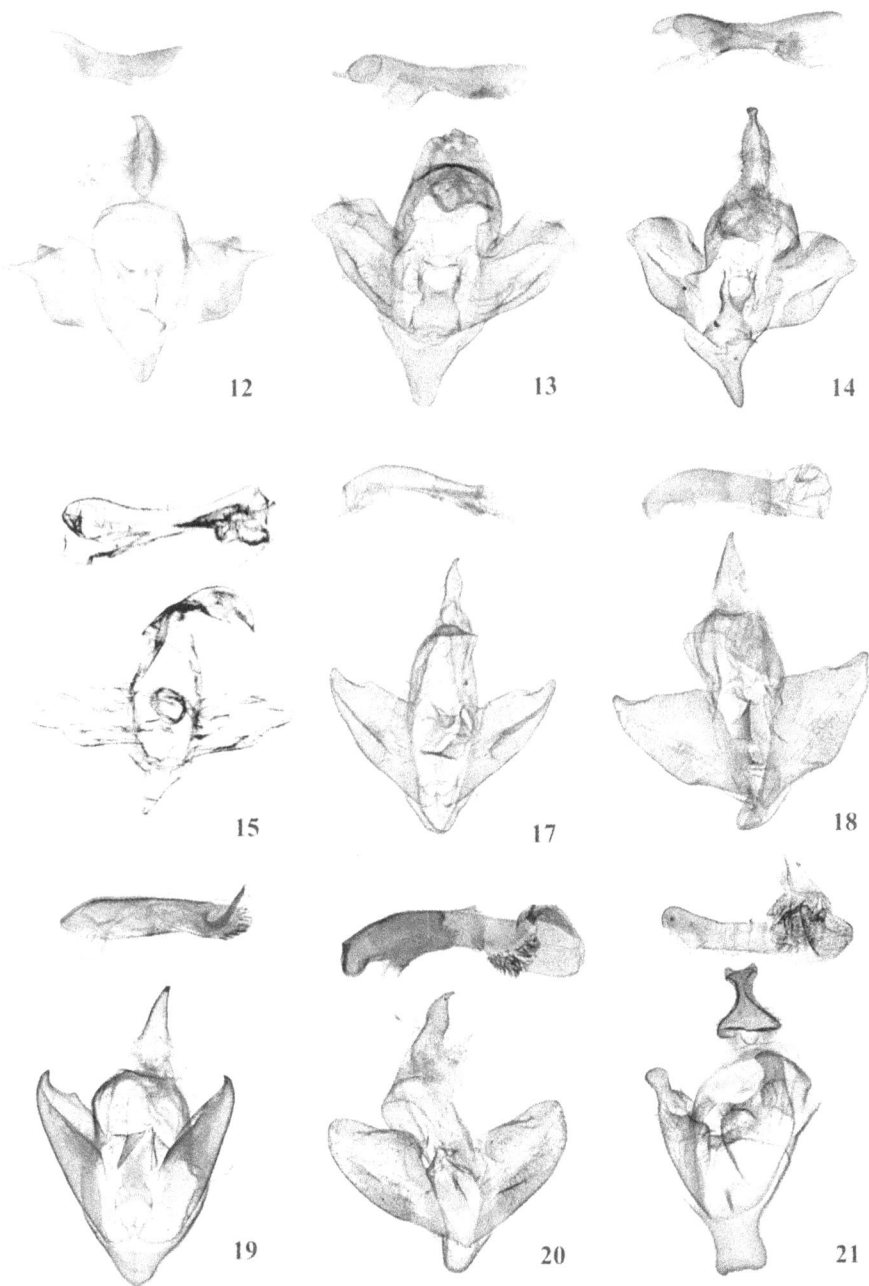

Plate 2

12. *Automolis crassa* (FELDER, 1874); RSA, "Stettyn's Kloof, Worcester Distr."; 17.10.1966; coll. TMSA. - **13.** *Automolis incensa* (WALKER, 1864); RSA, "Steynsburg"; 20.11.1965; coll. TMSA. - **14.** *Automolis meteus* (STOLL, 1780-82); RSA, "Blaney"; 03.1946; coll. ZSM. - **15.** *Automolis pallida* (HAMPSON, 1901); holotype of *subrosea* KIRIAKOFF, 1957; coll. BMNH. - **17.** *Balacra (Balacra) caeruleifascia* WALKER, 1856; holotype of *inflammata* HAMPSON, 1914; coll. BMNH. - **18.** *Balacra (Balacra) guillemei* (OBERTHÜR, 1911); DRC, "Ht. Katanga, Tshituru", coll. MHNG. - **19.** *Balacra (Balacra) nigripennis* (AURIVILLIUS, 1904); holotype of *aurivilliusi* KIRIAKOFF, 1957; coll. RMCA. - **20.** *Balacra (Balacra) rattrayi* (ROTHSCHILD, 1910); Rwanda, Kigali; coll. ZMUC. - **21.** *Balacra (Callobalacra) alberici* DUFRANE, 1945; DRC, Uele, Paulis; coll. RMCA.

127

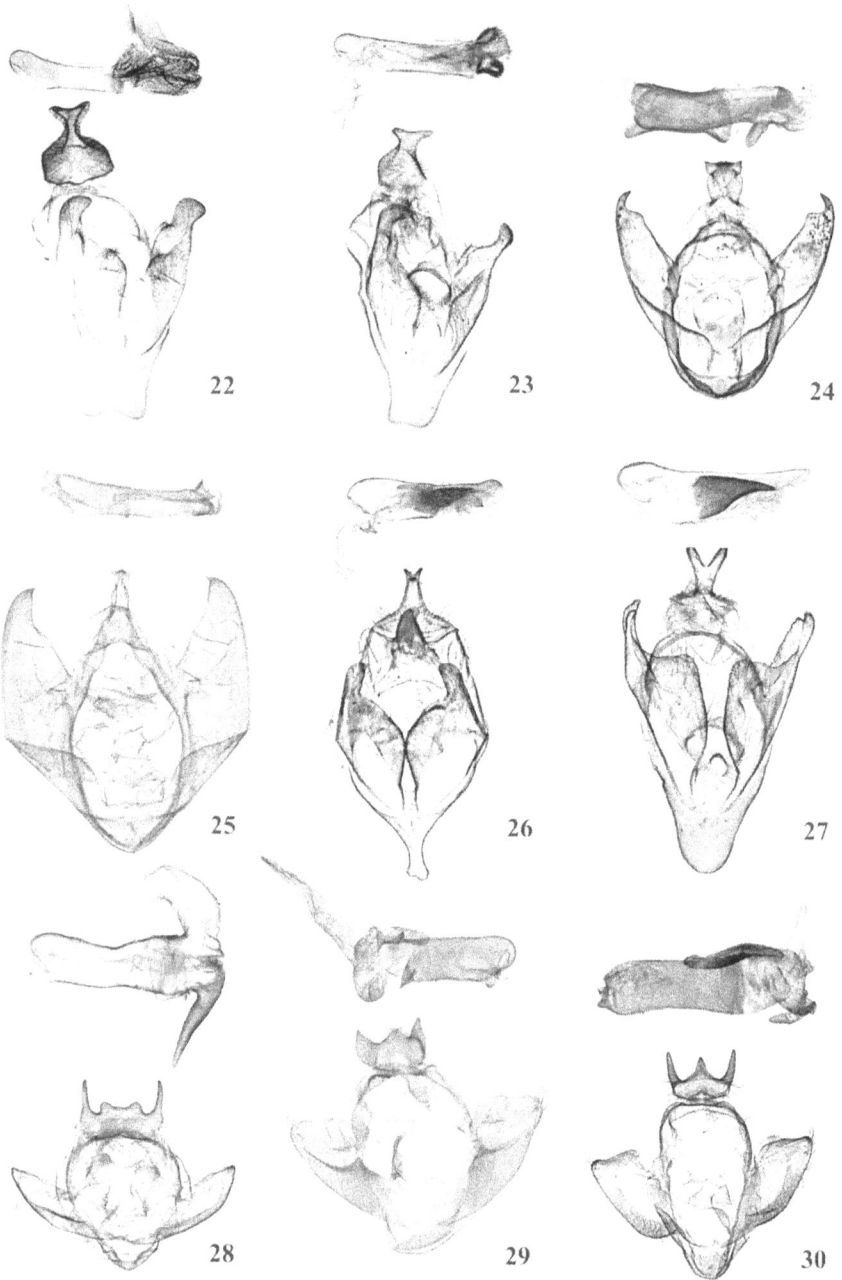

Plate 3

22. *Balacra (Callobalacra) jaensis* BETHUNE-BAKER, 1927, Cameroon, Eloumden; coll. RMCA. - 23. *Balacra (Callobalacra) rubrostriata* (AURIVILLIUS, 1898); Burundi, Gitega; coll. RMCA. - 24. *Balacra (Compsochromia) compsa* (JORDAN, 1904); Rwanda, Butare; coll. RMCA. - 25. *Balacra (Compsochromia) diaphana* KIRIAKOFF, 1957; holotype; coll. BMNH. - 26. *Balacra (Daphaenisca) affinis* (ROTHSCHILD, 1910); Sankuru, Dimbelenge; coll. RMCA. - 27. *Balacra (Daphaenisca) daphaena* (HAMPSON, 1898); Nigeria, Port Harcourt; coll. BMNH. - 28. *Balacra (Heronina) herona* (DRUCE, 1887); Nigeria, Bendel State, Okomu Forest Res.; coll. ISEA. - 29. *Balacra (Lamprobalacra) elegans* AURIVILLIUS, 1892; Nigeria, Bendel State, Okomu Forest Res.; 20.10.1984; coll. ISEA. - 30. *Balacra (Lamprobalacra) furva* HAMPSON, 1911; Ivory Coast, Foret classee de la Bossematie, 20 km S ad Abengourou; 02.1996; coll. ISEA.

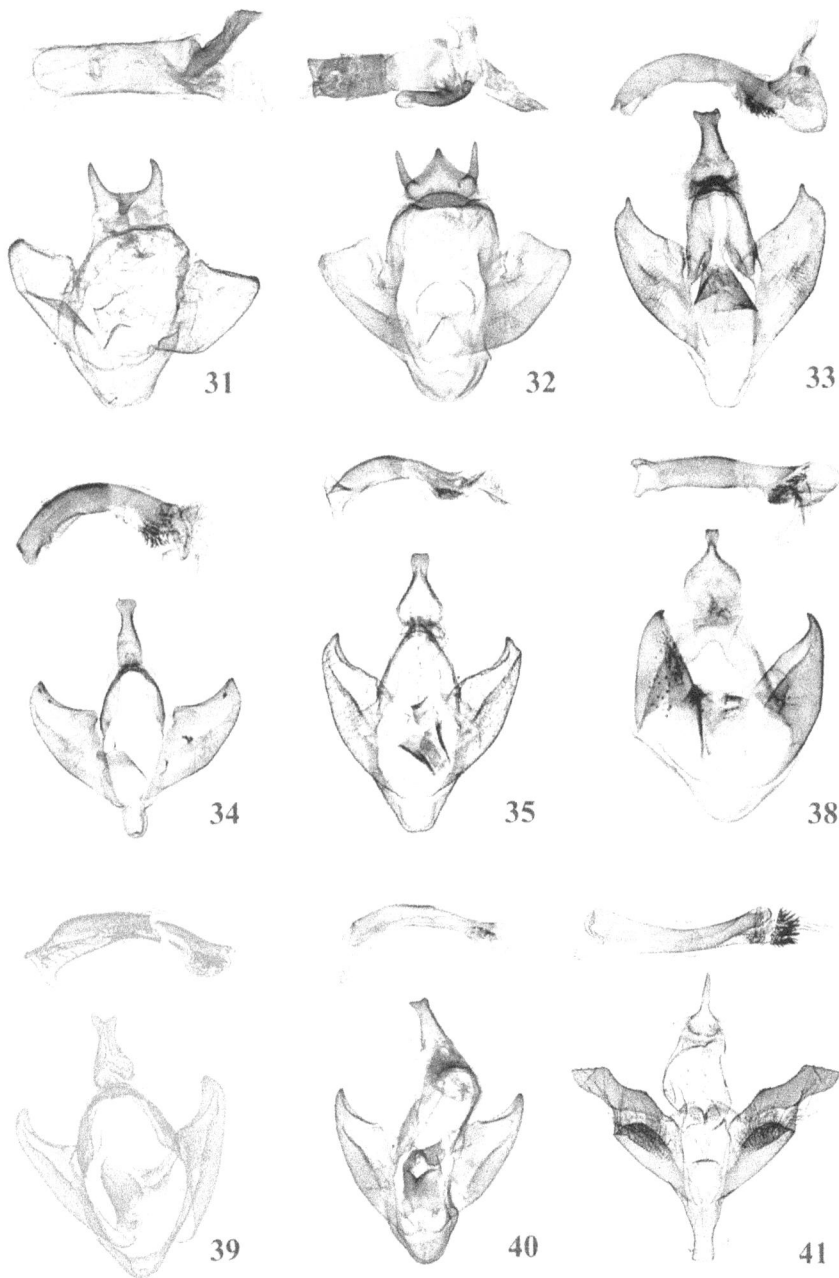

Plate 4

31. *Balacra (Lamprobalacra) pulchra* Aurivillius, 1892; Nigeria, Owenna; 20.07.1957; coll. ZMUC. - **32.** *Balacra (Lamprobalacra) rubricincta* Holland, 1893; Nigeria, Owenna; 22.07.1957; coll. ISEA. - **33.** *Balacra (Pseudapiconoma) basilewskyi* Kiriakoff, 1953; DRC, Uele, Paulis; 22.10.1956; coll. RMCA. - **34.** *Balacra (Pseudapiconoma) batesi* (Druce, 1910); DRC, Uele, Paulis; 15.03.1956; coll. RMCA. - **35.** *Balacra (Pseudapiconoma) flavimacula* Walker, 1856; holotype; coll. BMNH. - **38.** *Balacra (Pseudapiconoma) humphreyi* Rothschild, 1912; Ivory Coast, Foret classee de la Bossematie, 20 km S ad Abengourou; 02.1996; coll. ISEA. - **39.** *Balacra (Pseudapiconoma) monotonia* (Strand, 1912); holotype of *simplicior* Kiriakoff, 1957; coll. RMCA. - **40.** *Balacra (Pseudapiconoma) preussi* (Aurivillius, 1904); DRC, Lulonga, Merode; 01.1928.; coll. RMCA. - **41.** *Bergeria bourgognei* Kiriakoff, 1952; holotype of *fletcheri* Kiriakoff, 1957; coll. BMNH.

129

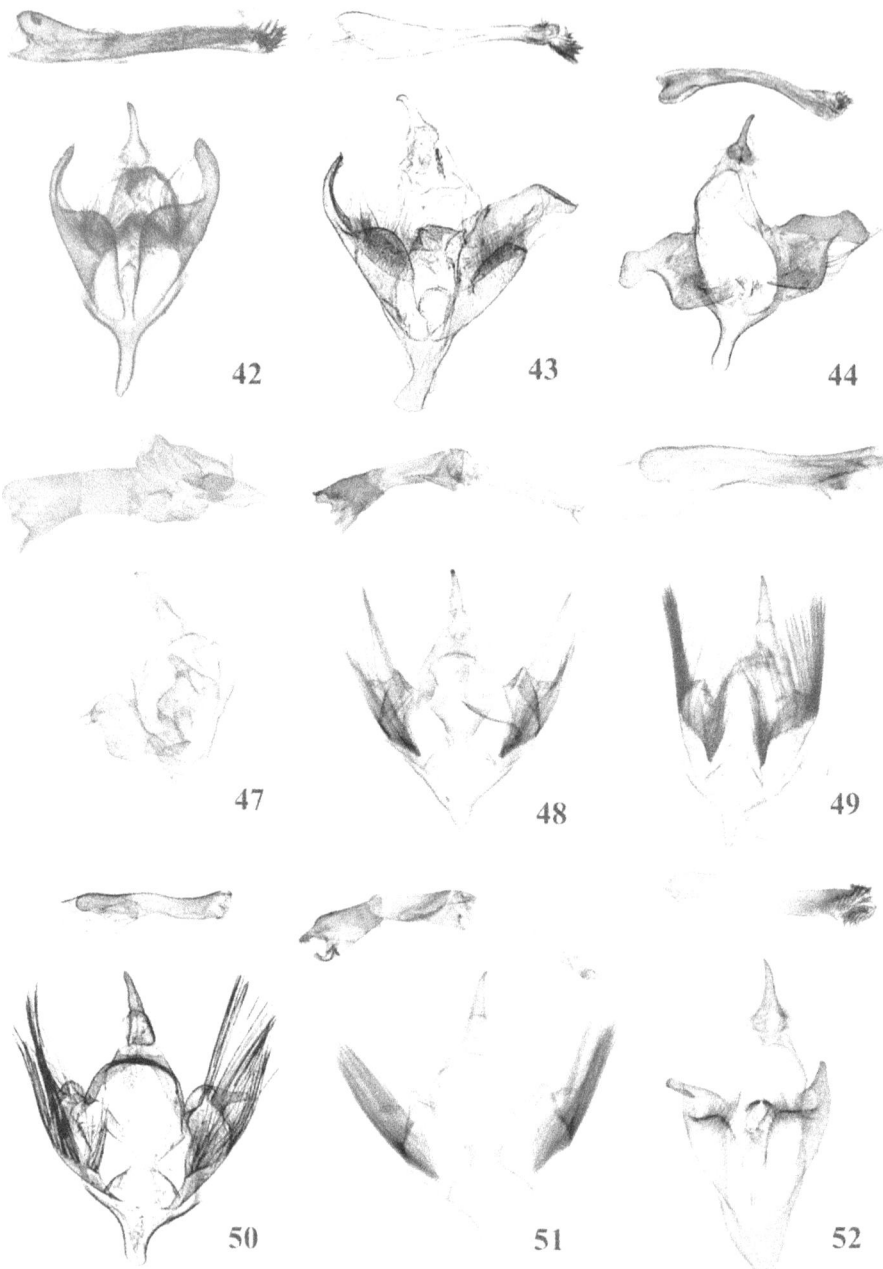

Plate 5

42. *Bergeria haematochrysia* Kɪʀɪᴀᴋᴏғғ, 1952; DRC, Sankuru, Dimbelenge; 13.03.1951; coll. RMCA. - **43.** *Bergeria octava* Kɪʀɪᴀᴋᴏғғ, 1961; holotype; coll. RMCA. - **44.** *Bergeria ornata* Kɪʀɪᴀᴋᴏғғ, 1959; DRC, Uele, Paulis; 19.06.1956; coll. RMCA. - **47.** *Cameroonia nigriceps* (Aᴜʀɪᴠɪʟʟɪᴜs, 1904); Cameroon, "Bitye, Ja River, 2000 ft"; 09-11.[19]11; coll.BMNH. - **48.** *Hippurarctia cinereoguttata* (Sᴛʀᴀɴᴅ, 1912); holotype; coll. ZMHB. - **49.** *Hippurarctia ferrigera* (Dʀᴜᴄᴇ, 1910); holotype of *vicina* Kɪʀɪᴀᴋᴏғғ, 1953; coll. RMCA. - **50.** *Hippurarctia judith* Kɪʀɪᴀᴋᴏғғ, 1959; holotype; DRC, Uele, Paulis; 26.03.1957; coll. RMCA. - **51.** *Hippurarctia taymansi* (Rᴏᴛʜsᴄʜɪʟᴅ, 1910); DRC, "W of Bukavu"; 17.05.1988; coll. ISEA. - **52.** *Lempkeella avellana* (Kɪʀɪᴀᴋᴏғғ, 1957); DRC, "Stanley Pool to Lukolele"; [18]94; coll. BMNH.

130

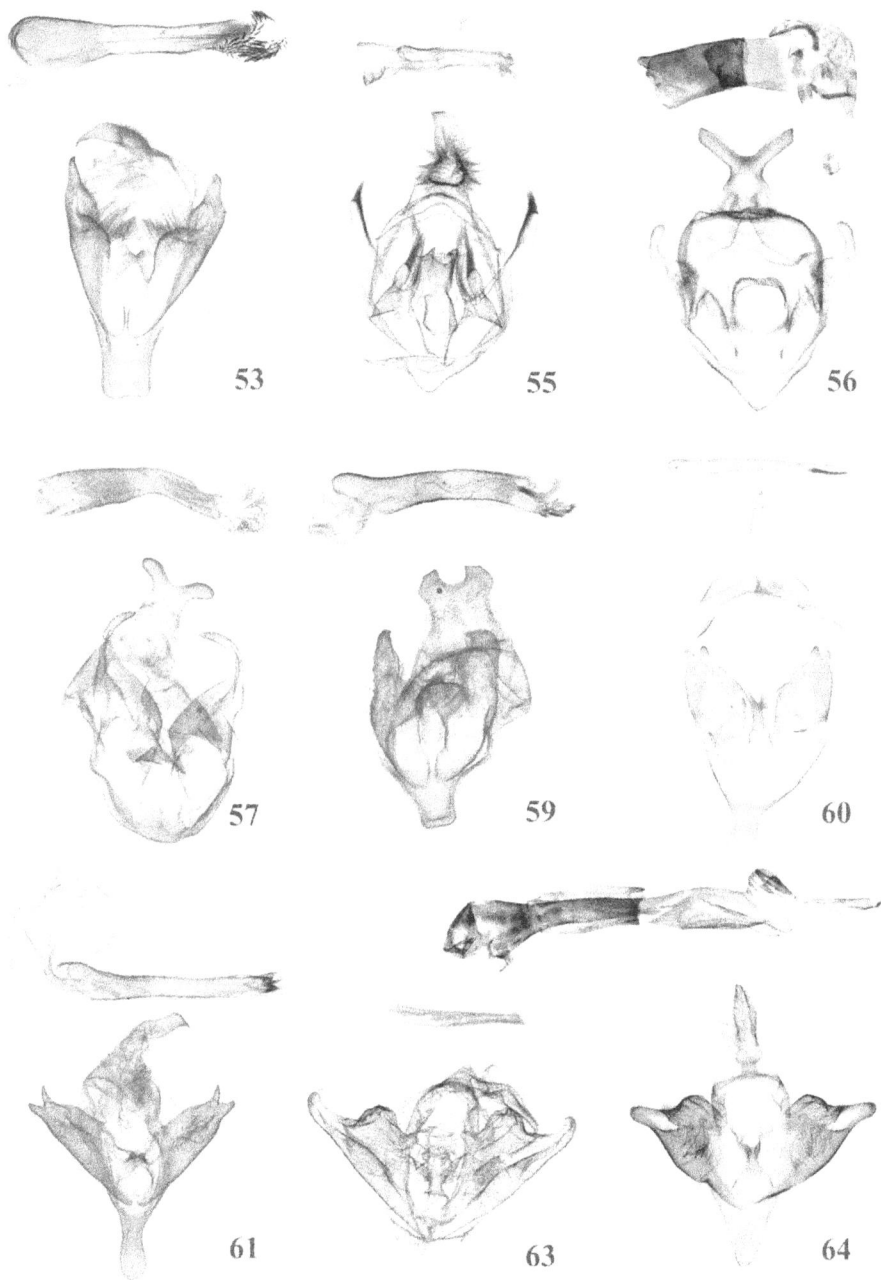

Plate 6

53. *Lempkeella dufranei* (KIRIAKOFF, 1952); DRC, "Tshuapa, Lukolela"; 8.09.1955; coll. RMCA. - **55.** *Mecistorhabdia haematoessa* (HOLLAND, 1893); holotype of *burgessi* KIRIAKOFF, 1957; coll. BMNH. - **56.** *Melisa croceipes* (AURIVILLIUS, 1892); Nigeria, Bendel State, Okomu F. Res.; 9.06.1986; coll. ISEA. - **57.** *Melisa diptera* (WALKER, 1854); Nigeria, Bendel State, Okomu F. Res.; 20.10.1984; coll. ISEA. - **59.** *Melisoides lobata* STRAND, 1912; DRC, Uele, Paulis; 10.09.1957; coll. RMCA. - **60.** *Metamicroptera christophi* PRZYBYŁOWICZ, 2005; holotype; coll. TMSA. - **61.** *Metamicroptera rotundata* HULSTAERT, 1923; DRC, "Elisabethville"; 10.12.1930; coll. RMCA. - **63.** *Metarctia (Collocalia) collocalia* KIRIAKOFF, 1957; holotype; coll. BMNH. - **64.** *Metarctia (Collocalia) debauchei* KIRIAKOFF, 1953; holotype; coll. RMCA.

131

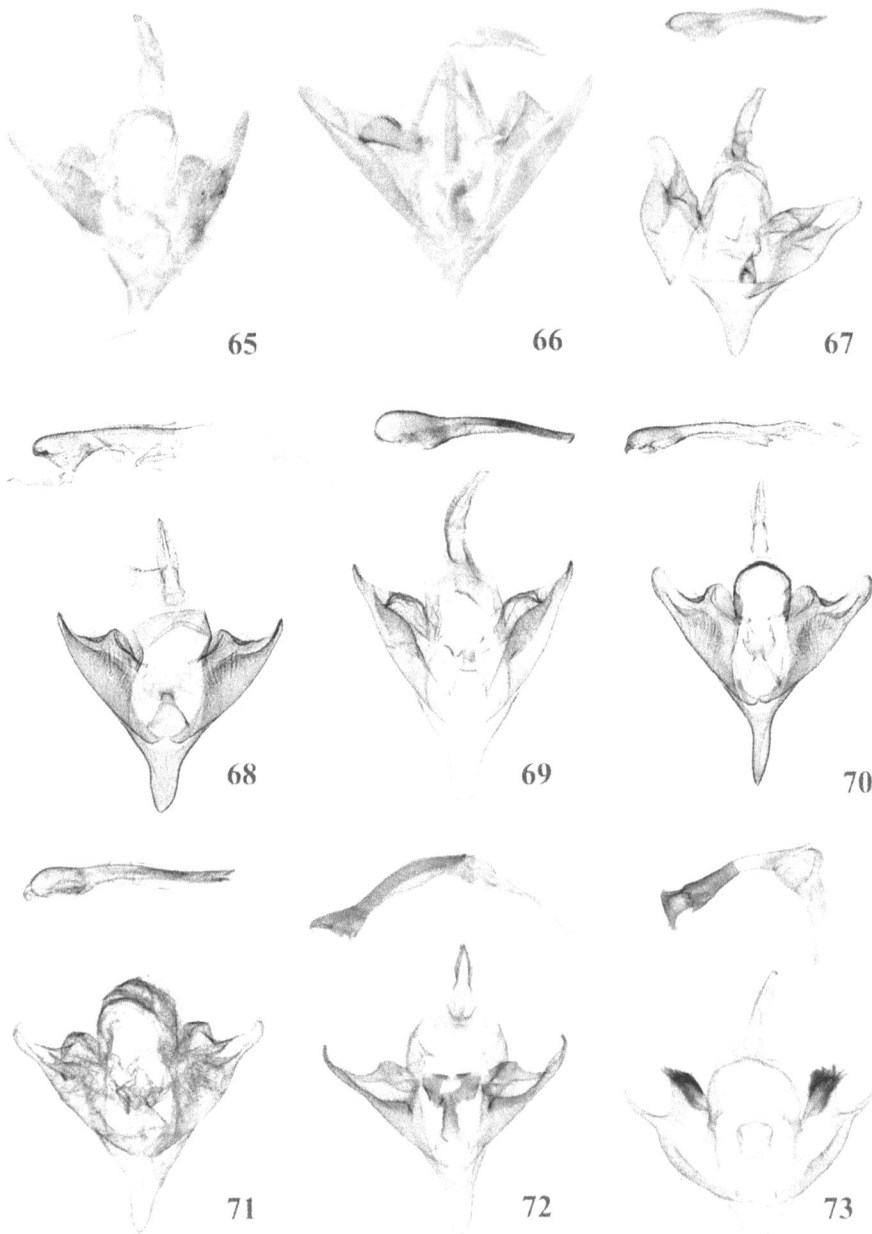

Plate 7

65. *Metarctia (Collocalia) dracoena* KIRIAKOFF, 1953; holotype; coll. RMCA. - **66.** *Metarctia (Collocalia) fuliginosa* KIRIAKOFF, 1953; holotype; coll. RMCA. - **67.** *Metarctia (Collocalia) jansei* KIRIAKOFF, 1957; holotype; coll. BMNH. - **68.** *Metarctia (Collocalia) olbrechtsi* KIRIAKOFF, 1953; paratype; coll. RMCA. - **69.** *Metarctia (Collocalia) pavlitzkae* KIRIAKOFF, 1961; holotype; coll. ZSM. - **70.** *Metarctia (Collocalia) seydeliana* KIRIAKOFF, 1953; paratype; coll. RMCA. - **71.** *Metarctia (Collocalia) tenebrosa* LE CERF, 1922; holotype of *margaretha* KIRIAKOFF, 1957; coll. BMNH. - **72.** *Metarctia (Hebena) cinnamomea* (WALLENGREN, 1860); RSA, "Ukhahlamba-Drakensberg Park, Cathedral Peak - Main Gate, 28°56'S 29°14'E; 1.12.2004; coll. ISEA. - **73.** *Metarctia (Hebena) henrardi* KIRIAKOFF, 1953; holotype; coll. RMCA.

132

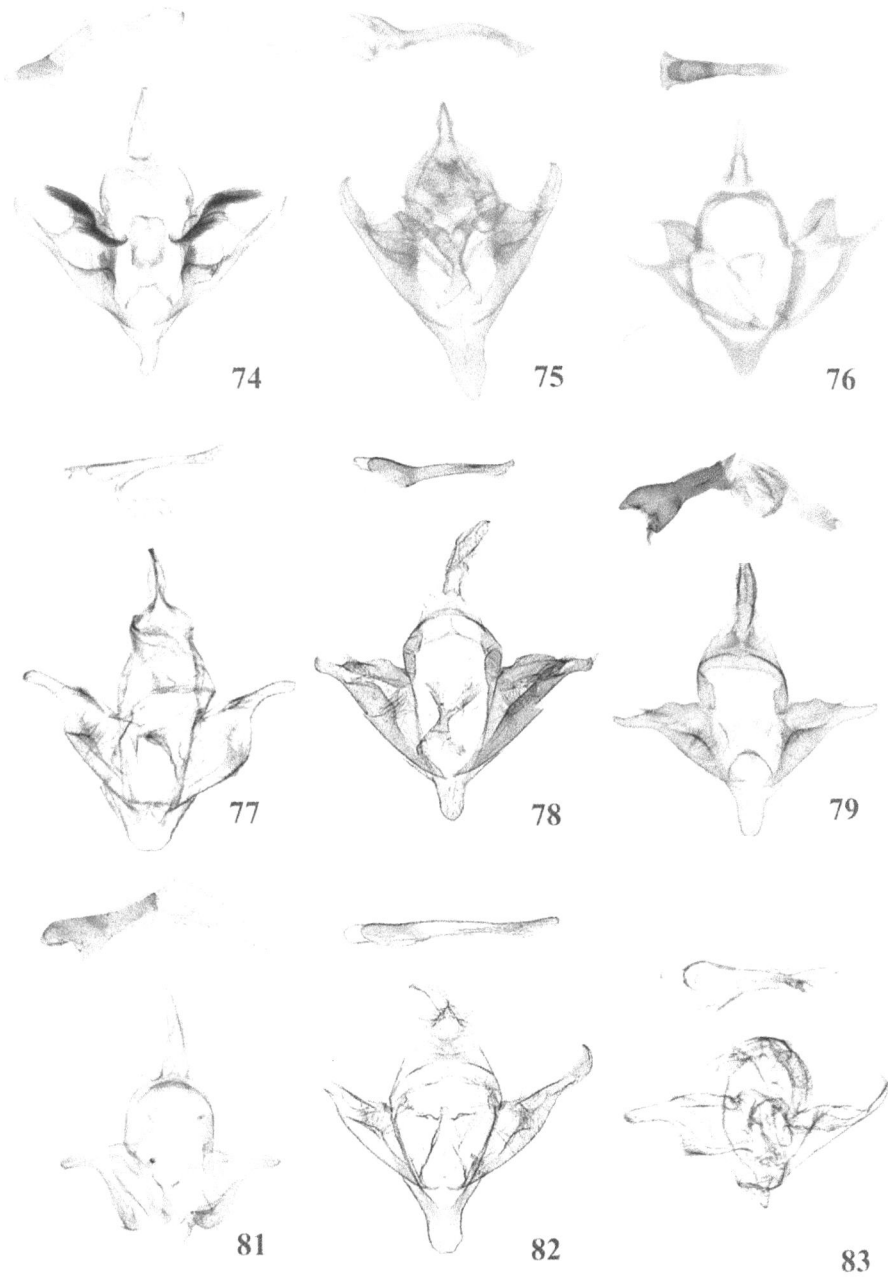

Plate 8

74. *Metarctia (Hebena) lateritia* (Herrich-Schaffer, 1850-1858); RSA; "Ukhahlamba-Drakensberg Park, Didima Camp, 28°56'S 29°14'E; 30.11.2004; coll. ISEA. - **75.** *Metarctia (Hebena) rubra* (Walker, 1856); syntype of *kelleni* Snellen, 1886; coll. RNHL. - **76.** *Metarctia (Hebena) subincarnata* Kiriakoff, 1954; holotype; coll. KBIN. - **77.** *Metarctia (Metarctia) alticola* Aurivillius, 1925; holotype of *rhodites* Kiriakoff, 1957; coll. BMNH. - **78.** *Metarctia (Metarctia) atrivenata* Kiriakoff, 1956; paratype; coll. RMCA. - **79.** *Metarctia (Metarctia) benitensis* Holland, 1893; syntype; coll. CMNH. - **81.** *Metarctia (Metarctia) burra* Schaus in Schaus & Clements, 1893; holotype; coll. AMNH. - **82.** *Metarctia (Metarctia) burungae* Debauche, 1942; holotype of *umbretta* Kiriakoff, 1963; coll. RMCA. - **83.** *Metarctia (Metarctia) carmel* Kiriakoff, 1957; holotype; coll. BMNH

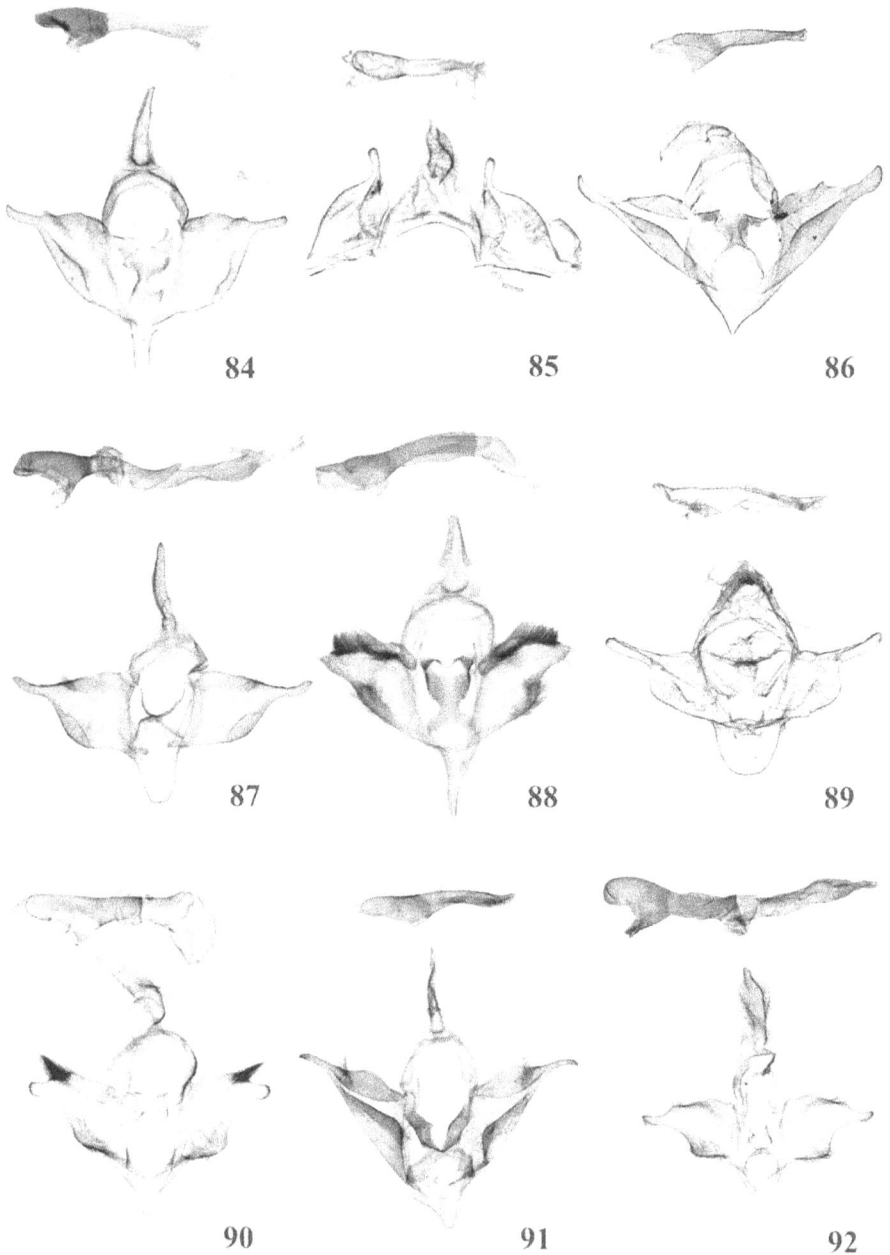

84. 85. 86.

84   85   86

87   88   89

90   91   92

Plate 9

84. *Metarctia (Metarctia) diversa* BETHUNE-BAKER, 1911; DRC, "Elisabethville, Ari W Kampf"; 17.03.37; coll. ZSM. - **85.** *Metarctia (Metarctia) fario* KIRIAKOFF, 1957; holotype; coll. BMNH. - **86.** *Metarctia (Metarctia) flaviciliata* HAMPSON, 1907; Uganda, "Katera"; 11.1953; coll. RMCA. - **87.** *Metarctia (Metarctia) flavicincta* AURIVILLIUS, 1900; holotype; coll. NHRS. - **88.** *Metarctia (Metarctia) flavivena* HAMPSON, 1901; Zambia, „Abercorn"; 9.12.1962; coll. RMCA. - **89.** *Metarctia (Metarctia) flora* KIRIAKOFF, 1957; holotype; coll. BMNH. - **90.** *Metarctia (Metarctia) fontainei* KIRIAKOFF, 1953; holotype; coll. RMCA. - **91.** *Metarctia (Metarctia) forsteri* KIRIAKOFF, 1955; holotype; coll. ZSM. - **92.** *Metarctia (Metarctia) fulvia* HAMPSON, 1901; Kenya, "Ulu"; 20.03.1958; coll. ISEA.

134

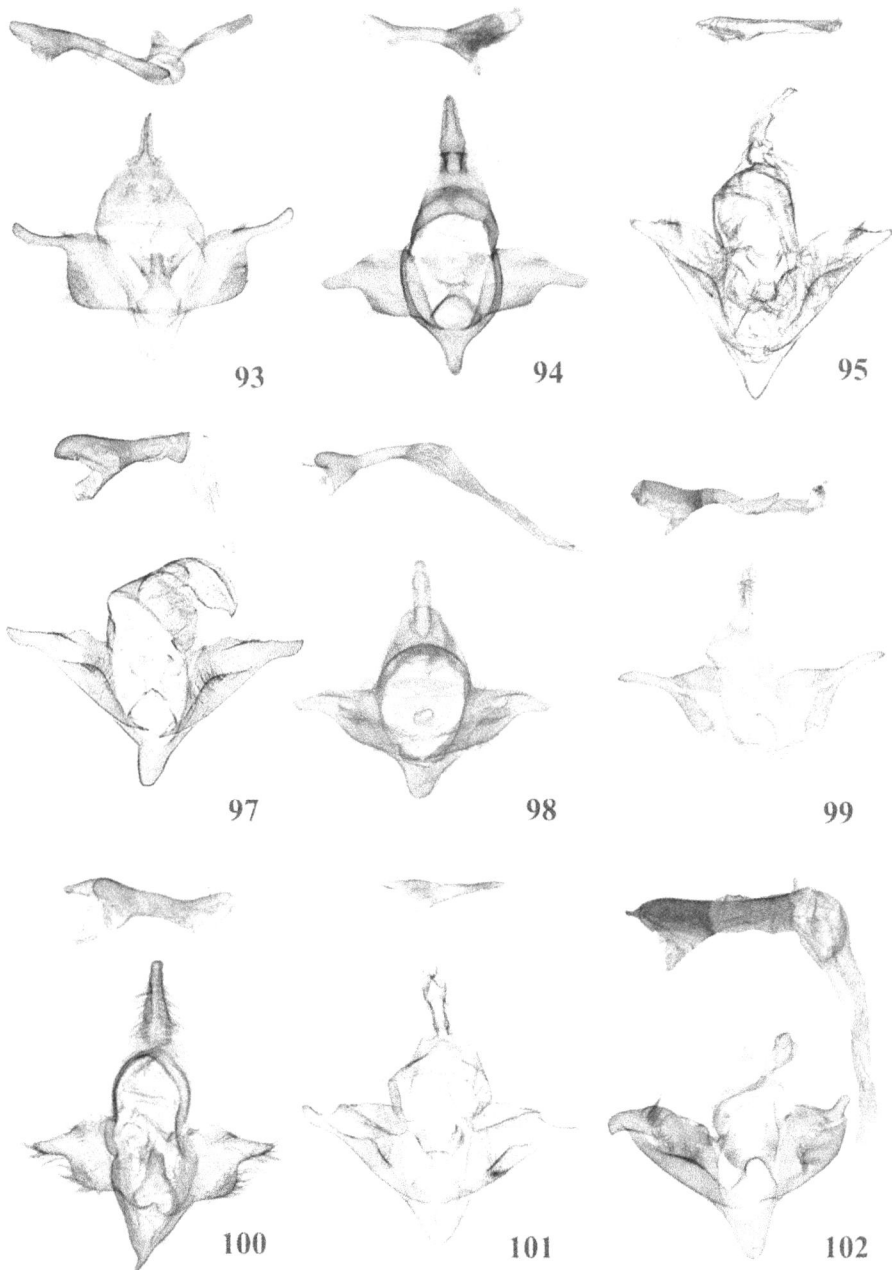

93 **94** **95**

97 **98** **99**

100 **101** **102**

Plate 10

93. *Metarctia (Metarctia) fusca* HAMPSON, 1901; [Uganda]; "Ruwenzori Range, Mahoma River, 6.700 ft"; 13-16.08.1952; coll. BMNH. - **94.** *Metarctia (Metarctia) galla* ROUGEOT, 1977: 82; holotype; coll. MNHN. - **95.** *Metarctia (Metarctia) haematricha* HAMPSON, 1905; holotype of *latipennis* KIRIAKOFF, 1957; coll. BMNH. - **97.** *Metarctia (Metarctia) hulstaertiana* KIRIAKOFF, 1953; paratype; coll. RMCA. - **98.** *Metarctia (Metarctia) inconspicua* HOLLAND, 1892; holotype; coll. USNM. - **99.** *Metarctia (Metarctia) johanna* (KIRIAKOFF, 1979); paratype; coll. ZSM. - **100.** *Metarctia (Metarctia) kumasina* STRAND, 1920; lectotype; coll. BMNH. - **101.** *Metarctia (Metarctia) lindemannae* KIRIAKOFF, 1961; holotype; coll. ZSM. - **102.** *Metarctia (Metarctia) longipalpis* HULSTAERT, 1923; holotype; coll. RMCA.

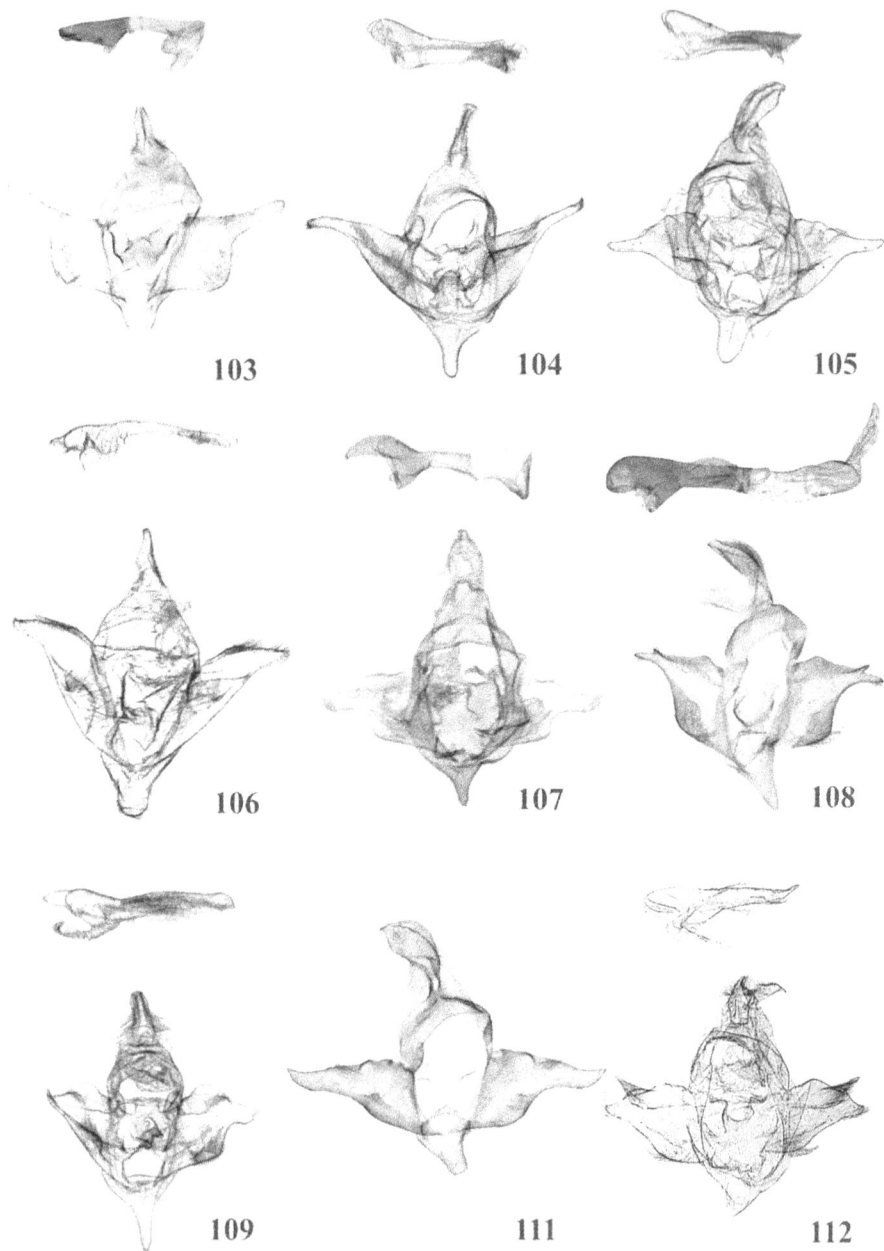

103   104   105

106   107   108

109   111   112

Plate 11

103. *Metarctia (Metarctia) lugubris* GAEDE, 1926; holotype; coll. ZMHB. - **104.** *Metarctia (Metarctia) maria* KIRIA-KOFF, 1957; holotype; coll. BMNH. - **105.** *Metarctia (Metarctia) metaleuca* HAMPSON, 1914; holotype; coll. BMNH. - **106.** *Metarctia (Metarctia) morag* KIRIAKOFF, 1957; holotype; coll. BMNH. - **107.** *Metarctia (Metarctia) negusi* KIRIAKOFF, 1957; holotype; coll. BMNH. - **108.** *Metarctia (Metarctia) nigritarsis* BERIO, 1943; holotype; coll. MCSNG. - **109.** *Metarctia (Metarctia) noctis* DRUCE, 1910; holotype; coll. BMNH. - **111.** *Metarctia (Metarctia) paremphares* HOLLAND, 1893; syntype; coll. CMNH. - **112.** *Metarctia (Metarctia) paulis* KIRIAKOFF, 1961; holotype; coll. RMCA.

136

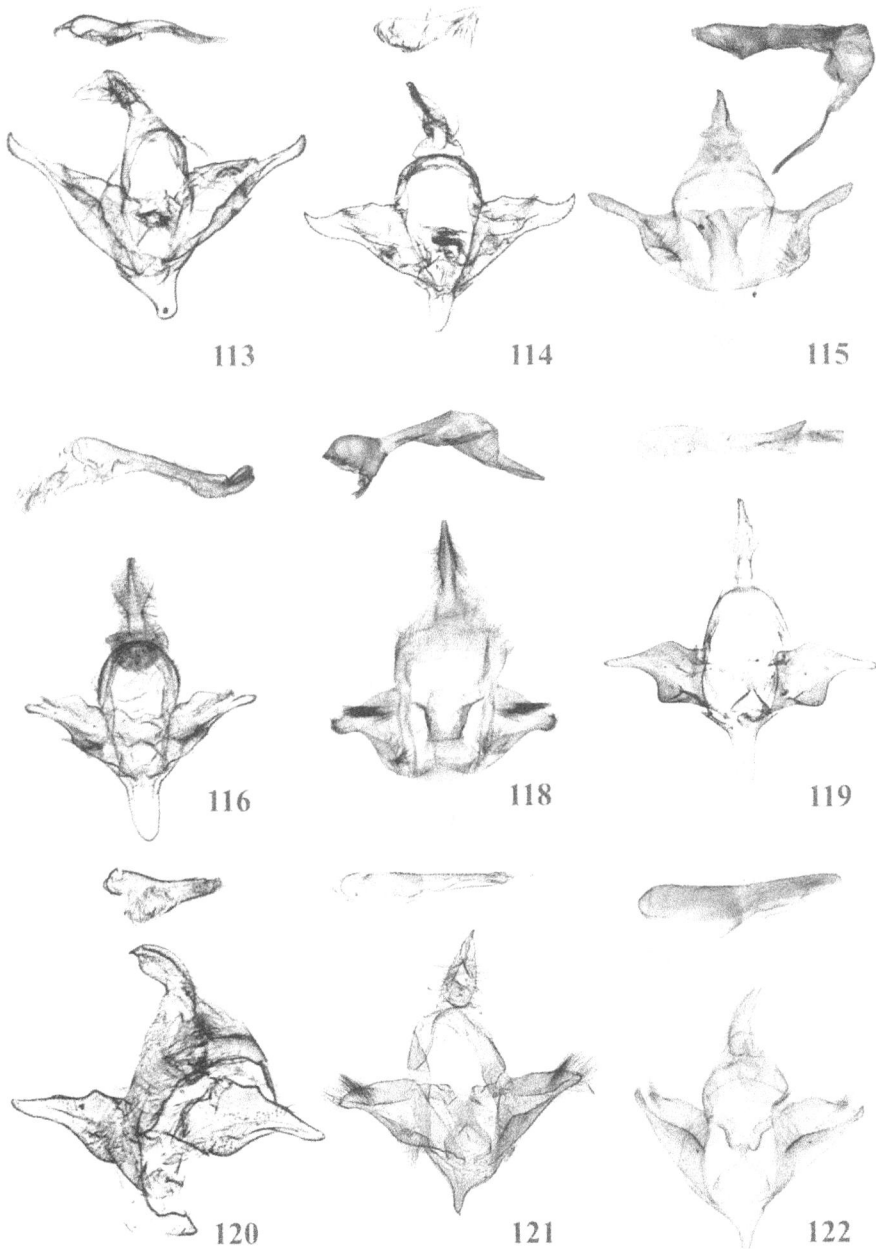

113   114   115

116   118   119

120   121   122

Plate 12

113. *Metarctia (Metarctia) phaeoptera* HAMPSON, 1909; holotype of *helga* KIRIAKOFF, 1957; coll. BMNH. - **114.** *Metarctia (Metarctia) priscilla* KIRIAKOFF, 1957; holotype; coll. BMNH. - **115.** *Metarctia (Metarctia) pulverea* HAMPSON, 1907; syntype of *bipuncta* JOICEY & TALBOT, 1924; coll. BMNH. - **116.** *Metarctia (Metarctia) pumila* HAMPSON, 1909; syntype; coll. BMNH. - **118.** *Metarctia (Metarctia) rufescens* WALKER, 1855; holotype of *maculifera* WALLENGREN, 1860; coll. NHRS. - **119.** *Metarctia (Metarctia) saalfeldi* KIRIAKOFF, 1960; holotype; coll. ZSM. - **120.** *Metarctia (Metarctia) salmonea* KIRIAKOFF, 1957; holotype; coll. BMNH. - **121.** *Metarctia (Metarctia) sarcosoma* HAMPSON, 1901; Kenya, "Kakamega"; 06.1950; coll. RMCA. - **122.** *Metarctia (Metarctia) sheljuzhkoi* KIRIAKOFF, 1961; holotype; coll. ZSM.

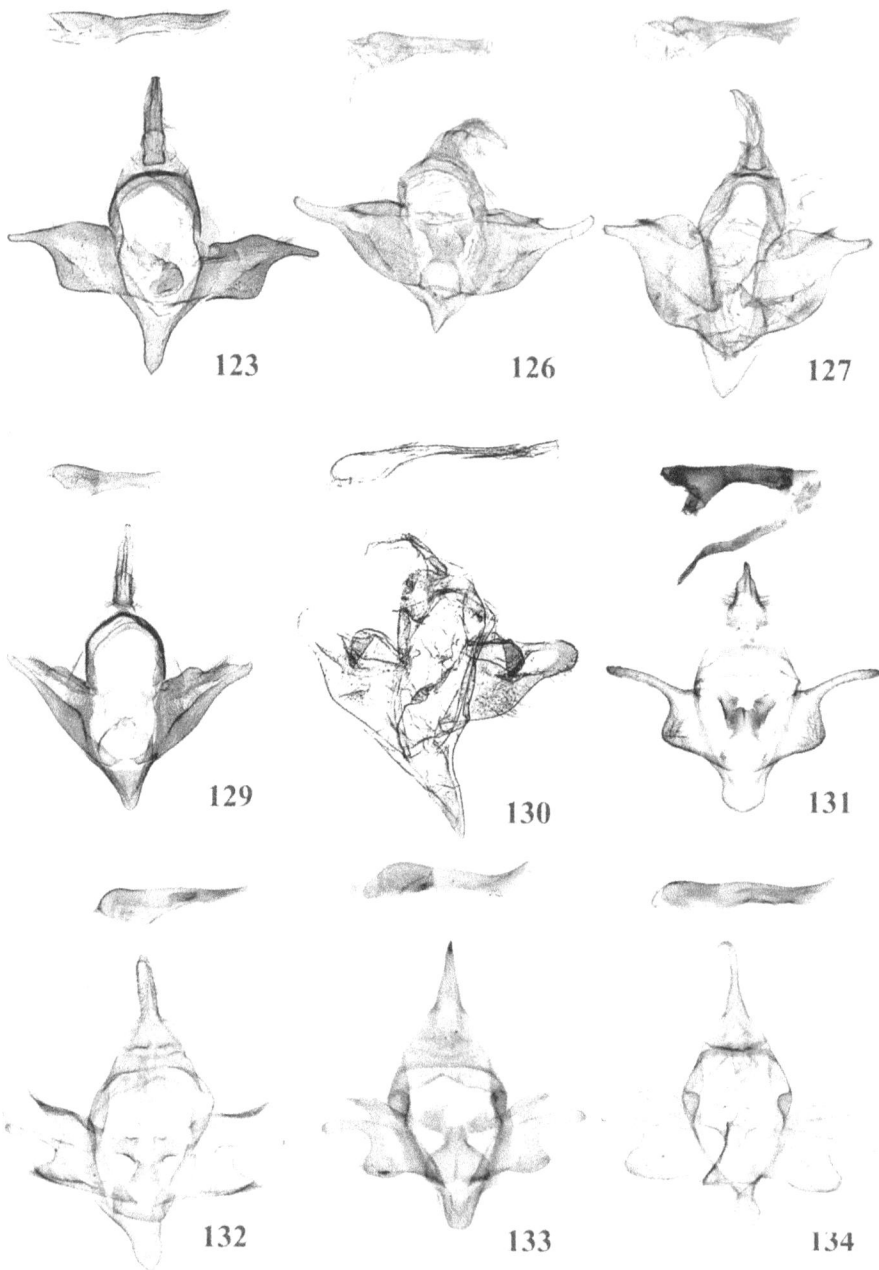

Plate 13

123. *Metarctia (Metarctia) subpallens* KIRIAKOFF, 1956; holotype; coll. RMCA. - 126. *Metarctia (Metarctia) tricolorana* WICH-GRAF, 1922; holotype; coll. BMNH. - 127. *Metarctia (Metarctia) unicolor* OBERTHÜR, 1880; holotype of *abyssinibia* KIRIAKOFF, 1957; coll. BMNH. - 129. *Metarctia (Metarctia) upembae* KIRIAKOFF, 1954; paratype; coll. RMCA. - 130. *Metarctia (Metarctia) venustissima* KIRIA-KOFF, 1961; holotype; coll. RMCA. - 131. *Metarctia (Metarctia) virgata* JOICEY & TALBOT, 1921; paratype of *wittei* DEBAUCHE, 1942; coll. RMCA. - 132. *Metarctia (Metarhodia) confederationis* KIRIAKOFF, 1961; holotype; coll. ZSM. - 133. *Metarctia (Metarhodia) epimela* (KIRIA-KOFF, 1979); holotype; coll. ZSM. - 134. *Metarctia (Metarhodia) heinrichi* KIRIAKOFF, 1961; holotype; coll. ZSM.

138

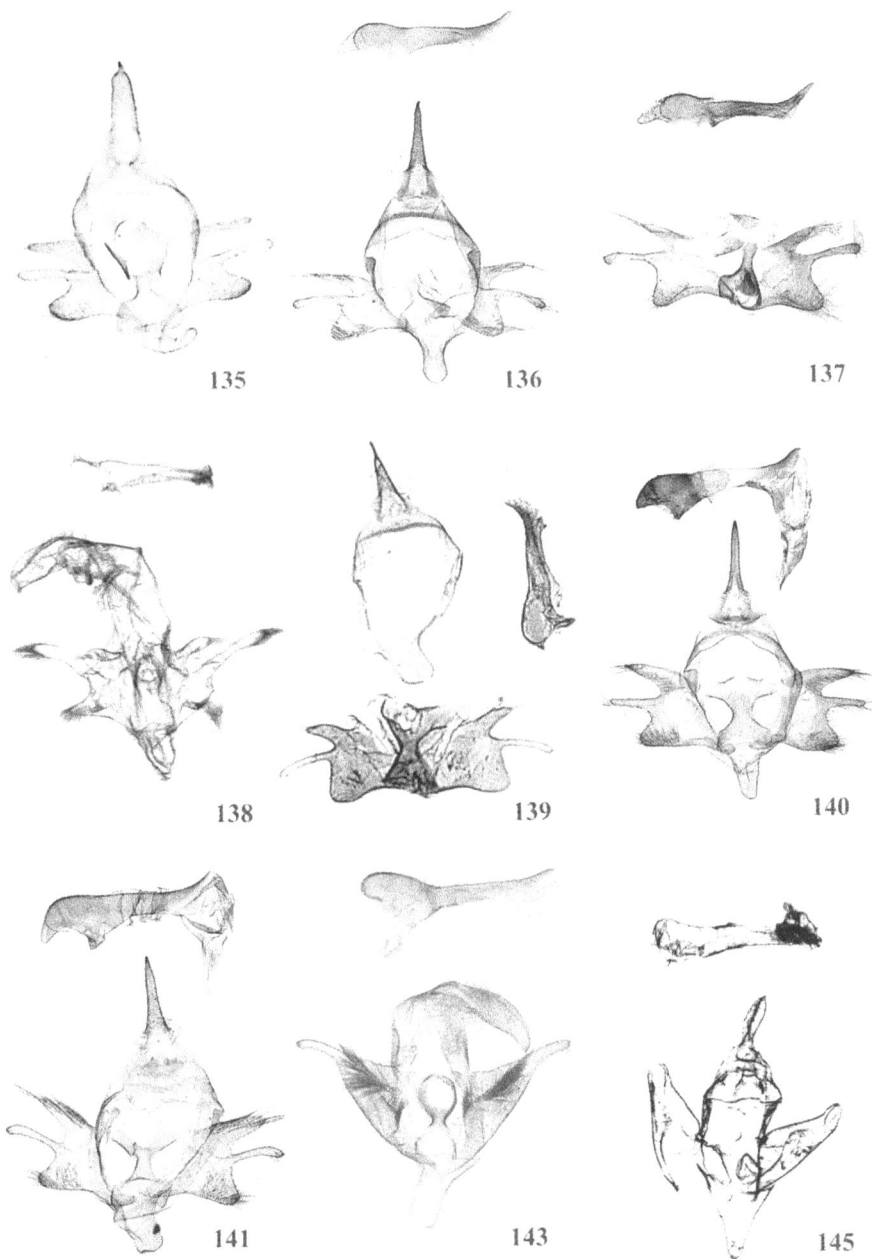

135   136   137

138   139   140

141   143   145

Plate 14

135. *Metarctia (Metarhodia) heringi* KIRIAKOFF, 1957; Botswana, Kasane Chobe; 26.11.1993; coll. ZMHB. - **136.** *Metarctia (Metarhodia) hypomela* KIRIAKOFF, 1956; holotype; coll. RMCA. - **137.** *Metarctia (Metarhodia) insignis* KIRIAKOFF, 1959; holotype; coll. RMCA. - **138.** *Metarctia (Metarhodia) jordani* KIRIAKOFF, 1957; holotype; coll. BMNH. - **139.** *Metarctia (Metarhodia) nigricornis* DEBAUCHE, 1942; holotype; coll. RMCA. - **140.** *Metarctia (Metarhodia) rubribasa* BETHUNE-BAKER, 1911; paratype of *deriemaeckeri* KIRIAKOFF, 1953; coll. RMCA. - **141.** *Metarctia (Metarhodia) rubripuncta* HAMPSON, 1898; paratype of *denisae* DUFRANE, 1945; coll. KBIN. - **143.** *Metarctia (Pinheyata) quinta* (KIRIAKOFF, 1973); Tanzania, "Mts Uluguru, Morogoro Campus Fac. Agric."; 05/06.1971; coll. RMCA. - **145.** *Metarctia (Thyretarctia) didyma* KIRIAKOFF, 1957; holotype; coll. BMNH.

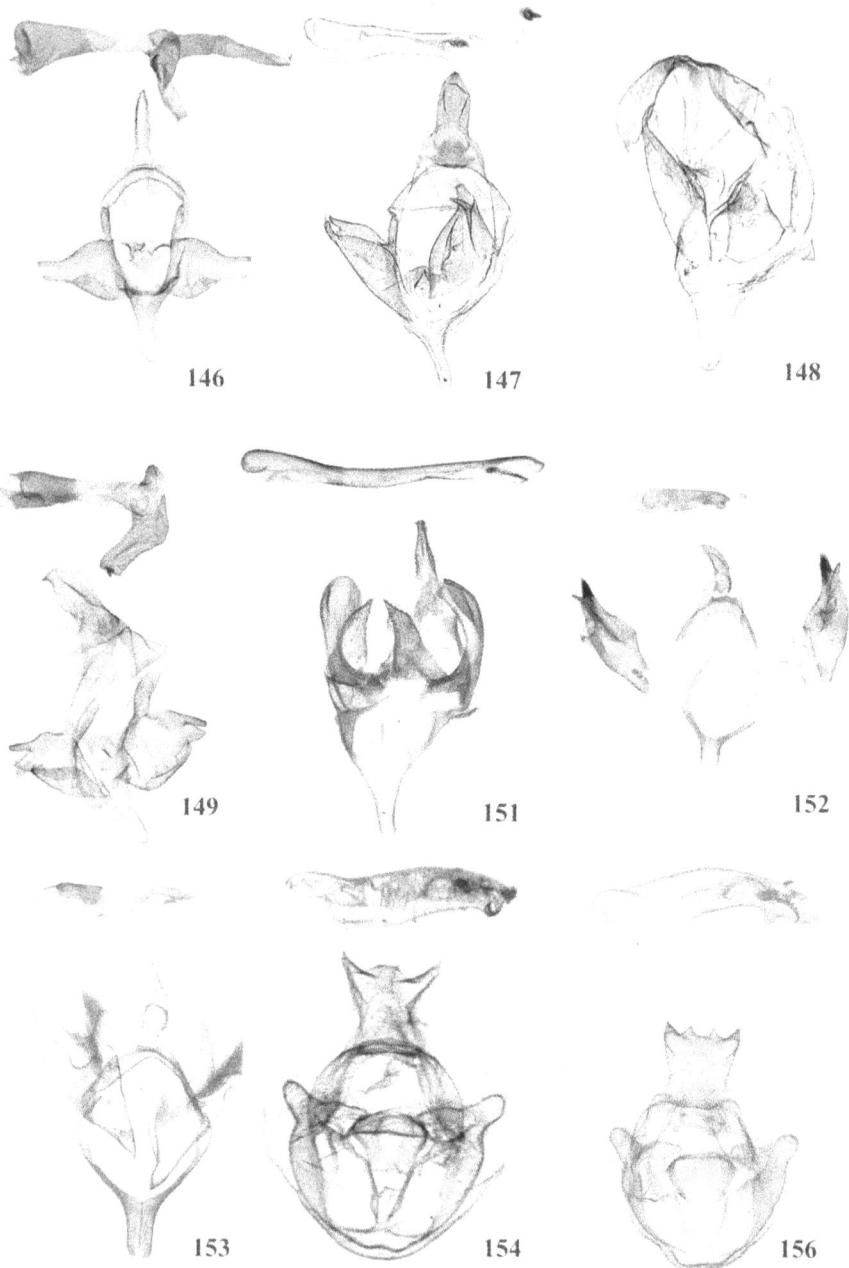

146   147   148

149   151   152

153   154   156

Plate 15

146. *Metarctia (Thyretarctia) haematica* HOLLAND, 1893; syntype; coll. CMNH. - **147.** *Metarctia (Thyretarctia) infausta* KIRIAKOFF, 1957; holotype; coll. RMCA. - **148.** *Metarctia (Thyretarctia) morosa* KIRIAKOFF, 1957; holotype; coll. RMCA. - **149.** *Metarctia (Thyretarctia) schoutedeni* KIRIAKOFF, 1953; holotype; coll. RMCA. - **151.** *Neophemula vitrina* (OBERTHÜR, 1909); holotype of ssp. *angolensis* KIRIAKOFF, 1957; coll. BMNH. - **152.** *Owambarctia owamboensis* KIRIAKOFF, 1957; holotype; coll. RMCA. - **153.** *Owambarctia unipuncta* KIRIAKOFF, 1973; holotype; coll. ZSM. - **154.** *Paramelisa dollmani* HAMPSON, 1920; lectotype; coll. BMNH. - **156.** *Paramelisa lophura* AURIVILLIUS, 1905; DRC, "Uele, Paulis"; 11.09.1957; coll. RMCA.

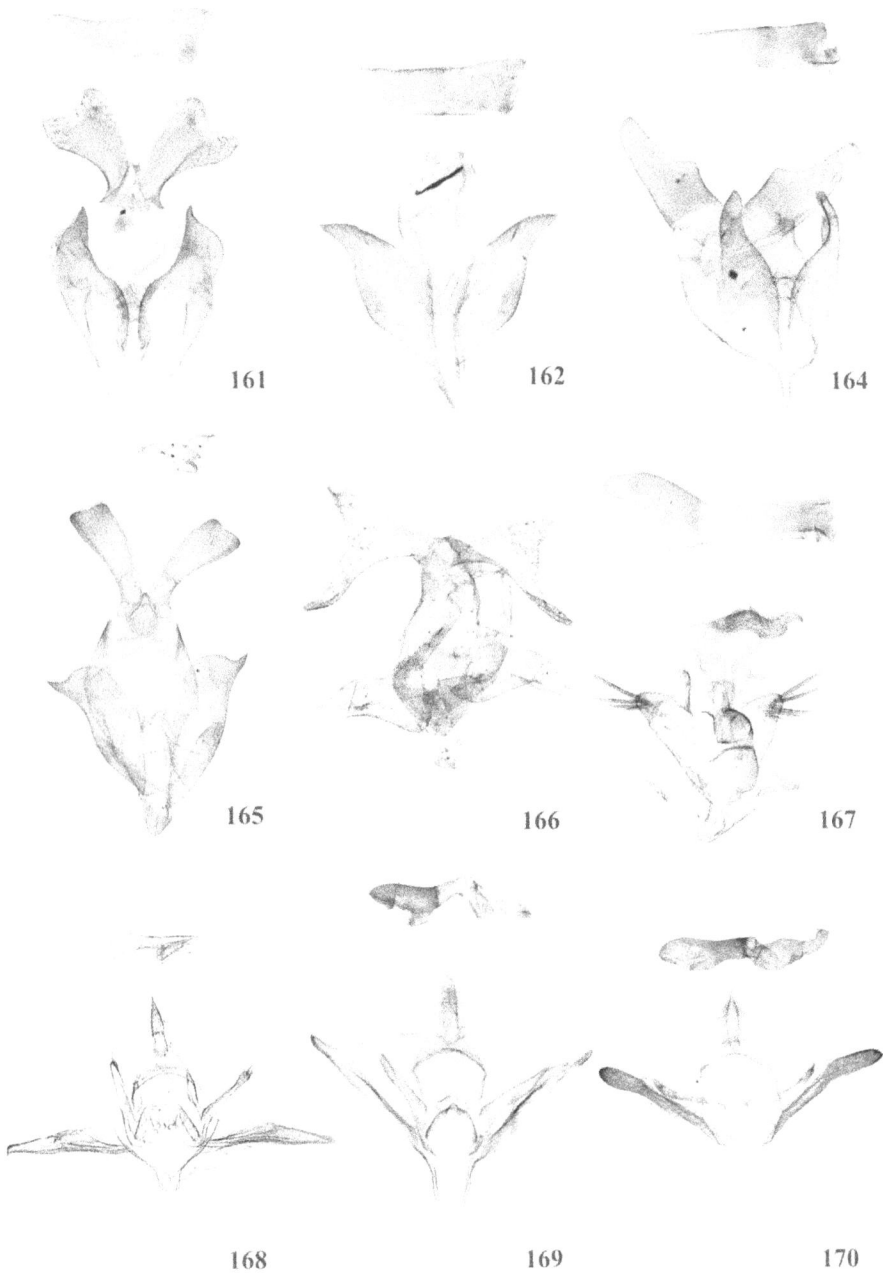

161   162   164

165   166   167

168   169   170

Plate 16

161. *Pseudothyretes erubescens* (HAMPSON 1901); Kenya, „Aberdares, Nat. Park Ruhuruini"; 2.04.2000; coll. RMCA. - **162.** *Pseudothyretes kamitugensis* (DUFRANE, 1945); Ivory Coast; 1913; coll. BMNH. - **164.** *Pseudothyretes nigrita* (KIRIAKOFF, 1961); DRC, "Kibali-Ituri, Ruwenzori"; 20.07.1952; coll. RMCA. - **165.** *Pseudothyretes perpusilla* (WALKER, 1856); Nigeria; coll. BMNH. - **166.** *Pseudothyretes rubicundula* (STRAND, 1912); holotype; coll. ZMHB. - **167.** *Rhabdomarctia rubrilineata* (BETHUNE-BAKER, 1911); DRC, "Lusambo"; 17.07.1949; coll. RMCA. - **168.** *Rhipidarctia (Elsitia) cinctella* (KIRIAKOFF, 1953); holotype of *strenua* KIRIAKOFF, 1957; coll. RMCA. - **169.** *Rhipidarctia (Elsitia) forsteri* (KIRIAKOFF, 1953); holotype of *punctulata* KIRIAKOFF, 1963; coll. RMCA. - **170.** *Rhipidarctia (Elsitia) lutea* (HOLLAND, 1893); holotype of *ghesquierei* KIRIAKOFF, 1953; coll. RMCA.

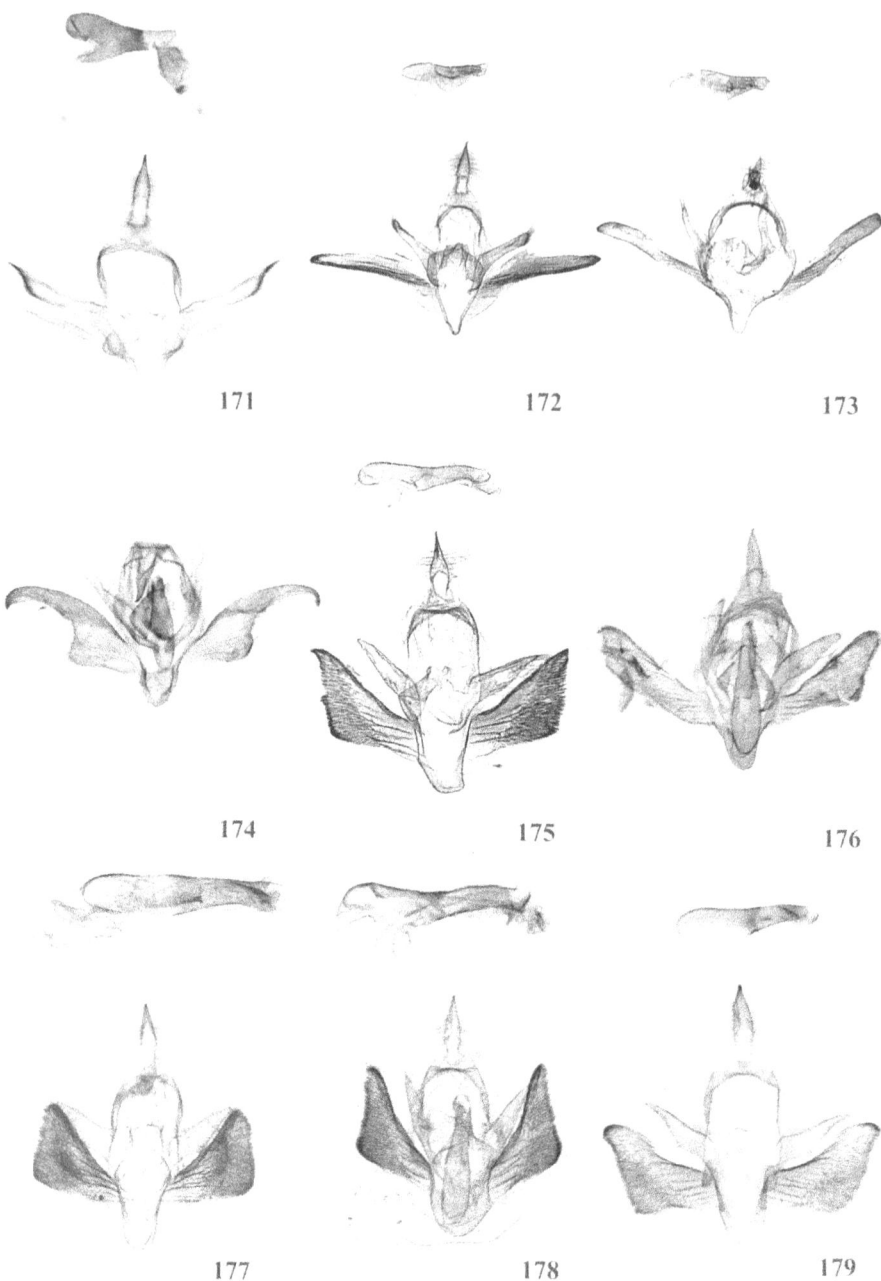

171   172   173

174   175   176

177   178   179

Plate 17

171. *Rhipidarctia (Elsitia) pareclecta* (HOLLAND, 1893); holotype; coll. CMNH. - **172.** *Rhipidarctia (Elsitia) saturata* KIRIAKOFF, 1957; DRC, "Sankuru, Katako-Kombe"; 10.01.1953; coll. RMCA. - **173.** *Rhipidarctia (Elsitia) subminiata* KIRIAKOFF, 1959; holotype; coll. RMCA. - **174.** *Rhipidarctia (Hemirhipidia) postrosea* (ROTHSCHILD, 1913); paratype of *danieli* KIRIAKOFF, 1955; coll. ZSM. - **175.** *Rhipidarctia (Rhipidarctia) aurora* KIRIAKOFF, 1957; paratype; coll. RMCA. - **176.** *Rhipidarctia (Rhipidarctia) conradti* (OBERTHUR, 1911); syntype; coll. BMNH. - **177.** *Rhipidarctia (Rhipidarctia) crameri* KIRIAKOFF, 1961; holotype; coll. ZSM. - **178.** *Rhipidarctia (Rhipidarctia) flaviceps* (HAMPSON, 1898); holotype of *rubrosuffusa* KIRIAKOFF, 1953; coll. RMCA. - **179.** *Rhipidarctia (Rhipidarctia) invaria* (WALKER, 1856); Nigeria, Anambra State, Nsukka F. Res.; 30.09.1982; coll. ISEA

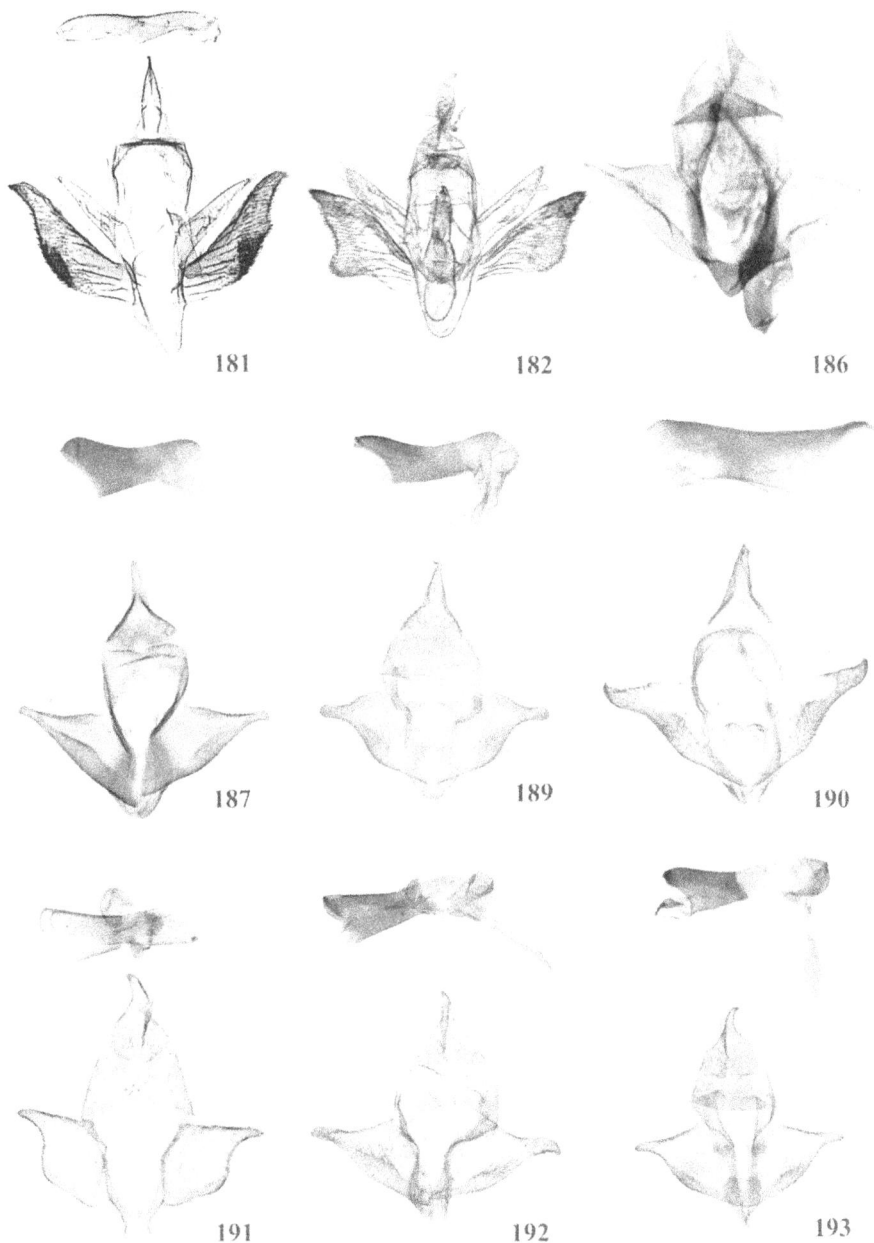

181   182   186

187   189   190

191   192   193

Plate 18

181. *Rhipidarctia (Rhipidarctia) rhodosoma* KIRIAKOFF, 1957; holotype; coll. RMCA. = **182.** *Rhipidarctia (Rhipidarctia) xenops* (KIRIA-KOFF, 1957); holotype; coll. BMNH. - **186.** *Thyretes buettikeri* WILTSHIRE, 1983; holotype; coll. NHMB. - **187.** *Thyretes caffra* WALLENGREN, 1863; RSA, "Pretoria"; 30.01.1978; coll. ISEA. - **189.** *Thyretes hippotes* (CRAMER, [1775-80]); RSA, "E Cape Prov., Katberg."; 14-26.11.1932; coll. BMNH. - **190.** *Thyretes montana* BOISDUVAL, 1847; RSA, "Natal, Karkloof"; 26.07.1916; coll. TMSA. - **191.** *Thyretes monteiroi* BUT-LER, 1876; holotype of *angolensis* GAEDE, 1926; coll. ZMHB. - **192.** *Thyretes negus* OBERTHÜR, 1878; Eritrea, "Dorfu"; 22.10.1938; coll. MC-SNG. - **193.** *Thyretes signivenis* HERING, 1937; paratype; coll. RMCA.

143

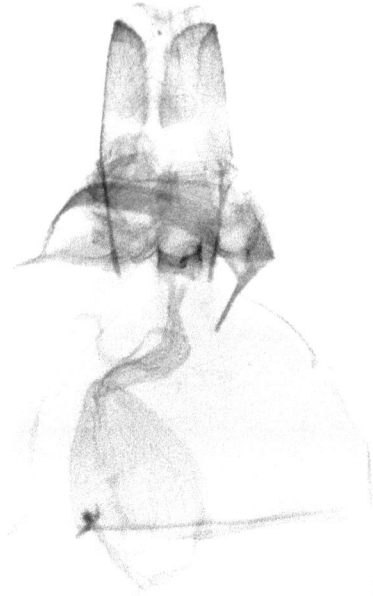

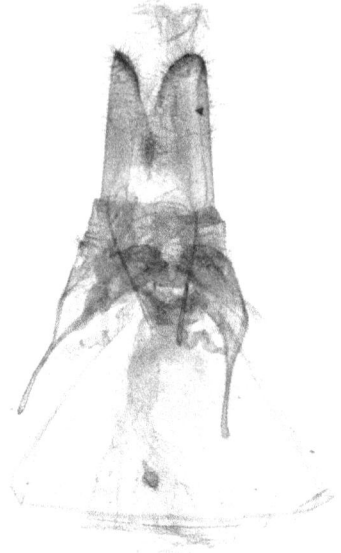

2

9

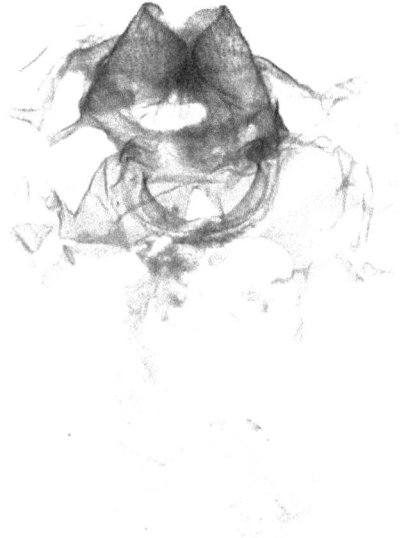

12

14

Plate 19

2. *Apisa (Apisa) canescens* WALKER, 1855; syntype of *cana* HOLLAND, 1893; coll. CMNH. - 9. *Apisa (Parapisa) subargentea* JOICEY & TALBOT, 1921; holotype; coll. BMNH. - 12. *Automolis crassa* (FELDER, 1874); [no locality]; coll. BMNH. - 14. *Automolis meteus* (STOLL, 1780-82); RSA, "New Denmark, Standerton"; 12.1942; coll. TMSA.

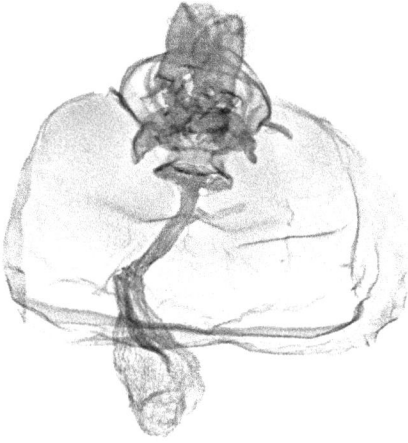

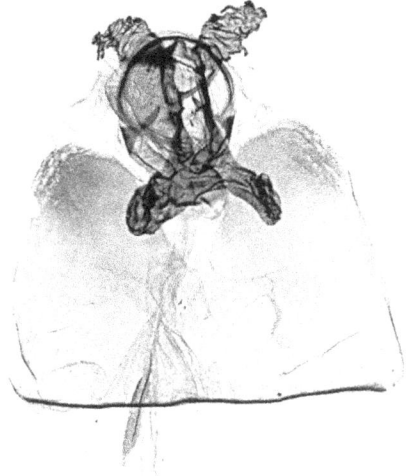

17

19

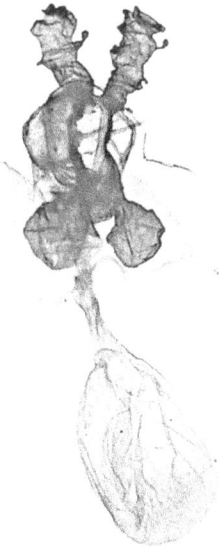

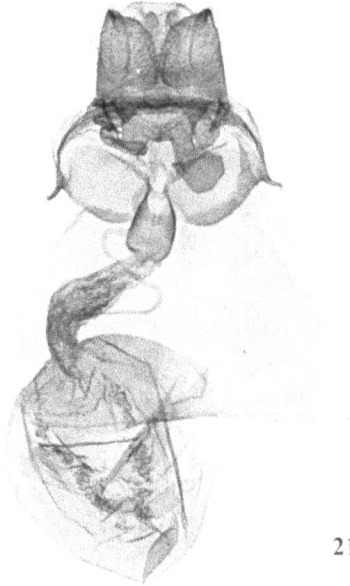

20

21

Plate 20

17. *Balacra (Balacra) caeruleifascia* WALKER, 1856; holotype; coll. BMNH. - **19.** *Balacra (Balacra) nigripennis* (AURIVILLIUS, 1904); DRC, Kambove, Katanga; 14.01.1907; coll. BMNH. - **20.** *Balacra (Balacra) rattrayi* (ROTHSCHILD, 1910); DRC, "Kissanyi"; coll. RMCA. - **21.** *Balacra (Callobalacra) alberici* DUFRANE, 1945; Zaire, Kalima, Maniema; 18.10.[19]54; coll. KBIN

145

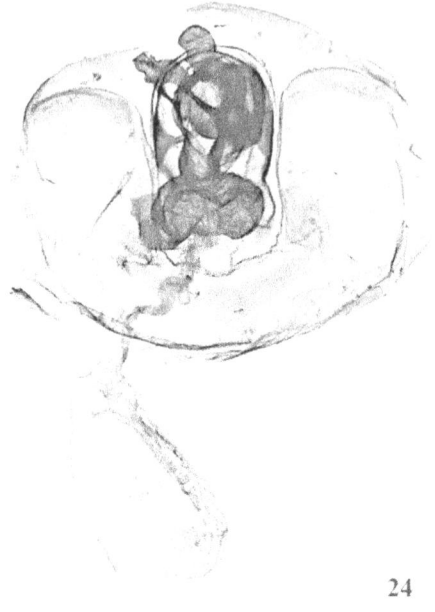

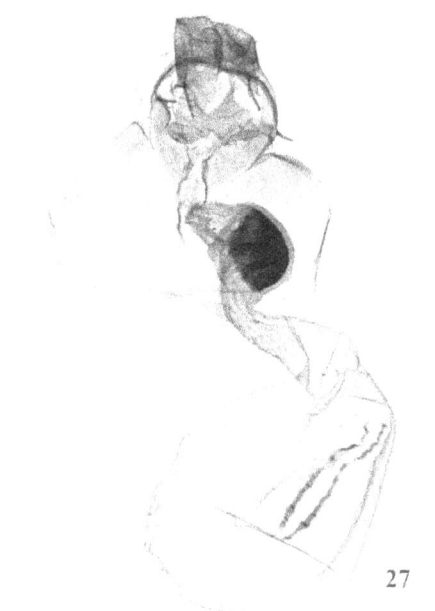

Plate 21

23. *Balacra (Callobalacra) rubrostriata* (Aurivillius, 1898); Zaire, Uele, Paulis; coll. RMCA. - **24.** *Balacra (Compsochromia) compsa* (Jordan, 1904); Uele, Dingila; coll. RMCA. - **26.** *Balacra (Daphaenisca) affinis* (Rothschild, 1910); Zaire, Eala; coll. RMCA. - **27.** *Balacra (Daphaenisca) daphaena* (Hampson, 1898); Nigeria, Port Harcourt; coll. BMNH.

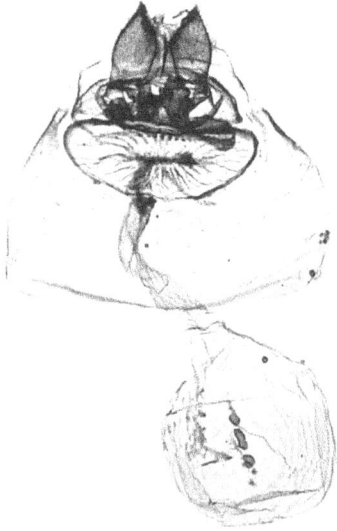

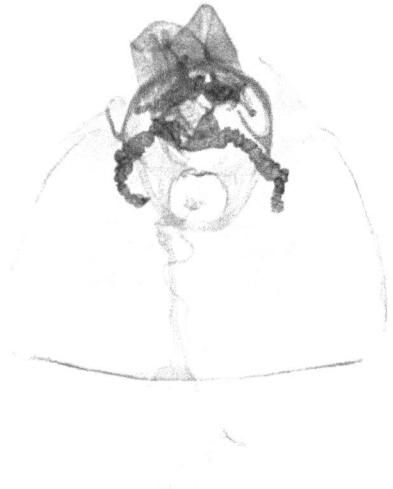

28

29

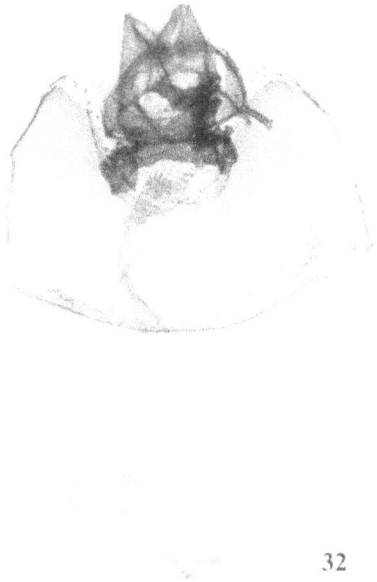

31

32

Plate 22

28. *Balacra (Heronina) herona* (DRUCE, 1887); Gold Coast, Bibianaha, 70 miles NW of Dimkwa; coll. BMNH. - **29.** *Balacra (Lamprobalacra) elegans* AURIVILLIUS, 1892; Uganda; Kyagive, Mulange, Mabera Forest; 04-08.[..]19; coll. BMNH. - **31.** *Balacra (Lamprobalacra) pulchra* AURIVILLIUS, 1892; Zaire, Uele, Paulis; 7.05.1960; coll. RMCA. - **32.** *Balacra (Lamprobalacra) rubricincta* HOLLAND, 1893; Zaire, Coquilhatville; 28.03.1925; coll. BMNH.

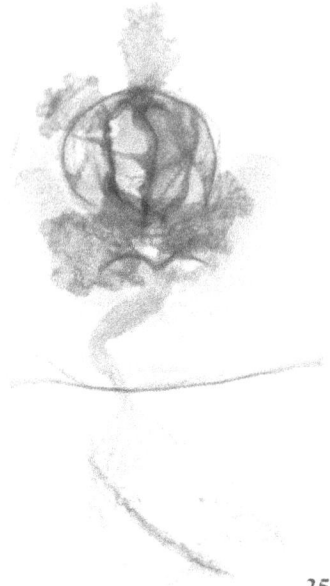

34

35

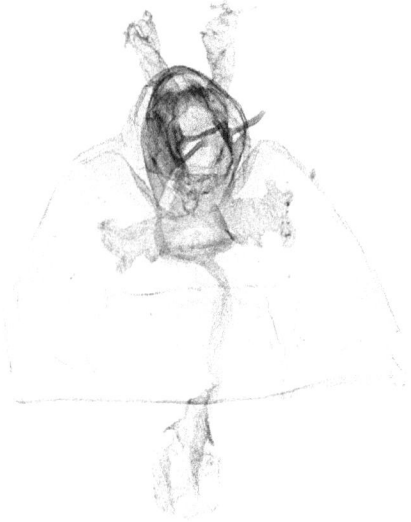

39

40

Plate 23

34. *Balacra* (*Pseudapiconoma*) *batesi* (DRUCE, 1910); Cameroon, "Bitje, Ja River"; coll. BMNH. - **35.** *Balacra* (*Pseudapiconoma*) *flavimacula* WALKER, 1856; Uganda, Entebbe; coll. BMNH. - **39.** *Balacra* (*Pseudapiconoma*) *monotonia* (STRAND, 1912); Cameroon, "Bitje, Ja River"; coll. BMNH. - **40.** *Balacra* (*Pseudapiconoma*) *preussi* (AURIVILLIUS, 1904); DRC, "Eala"; 14.11.1931; coll. RMCA.

42

44

45

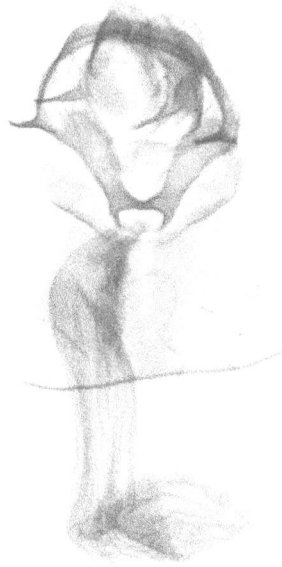

47

Plate 24

42. *Bergeria haematochrysia* KIRIAKOFF, 1952; paratype; coll. RMCA. - **44.** *Bergeria ornata* KIRIAKOFF, 1959; DRC, Uele, Paulis; 3.07.1956; coll. RMCA. - **45.** *Bergeria schoutedeni* KIRIAKOFF, 1952; DRC, Tshuapa, Bamanya; 23.09.1966; coll. RMCA. - **47.** *Cameroonia nigriceps* (AURIVILLIUS), 1904; holotype; coll. NHRS

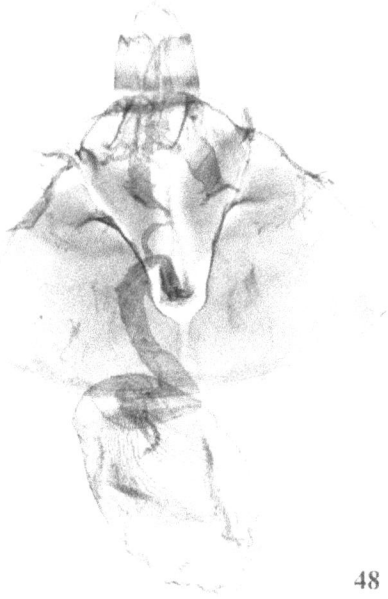
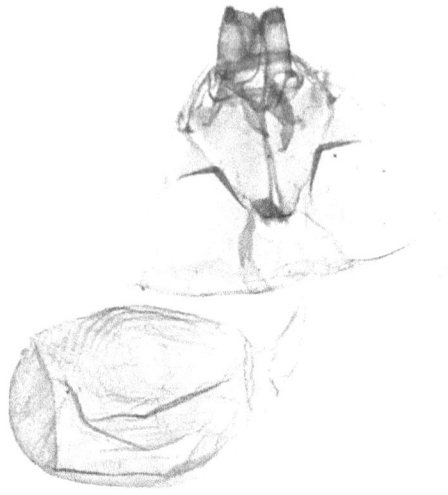

48

49

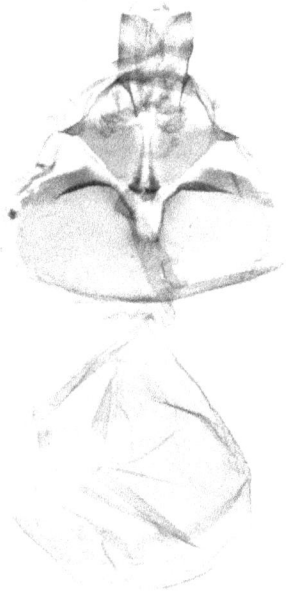
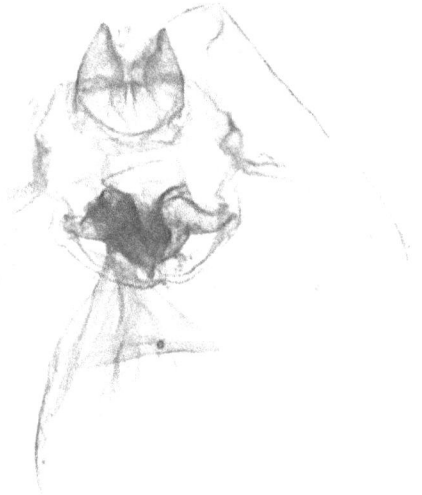

50

55

Plate 25

48. *Hippurarctia cinereoguttata* (STRAND, 1912); Congo, Lefinie res., Mbeokala forest; 8.01.1964; coll. ISEA. - **49.** *Hippurarctia ferrigera* (DRUCE, 1910); paratype of *vicina overlaeti* KIRIAKOFF, 1953; coll. RMCA. - **50.** *Hippurarctia judith* KIRIAKOFF, 1959; Nigeria, Bendel State, Okomu F. Res.; 20.01.1985; coll. ISEA. - **55.** *Mecistorhabdia haematoessa* (HOLLAND, 1893); DRC, Uele, Paulis; 30.07.1959; coll. RMCA.

56

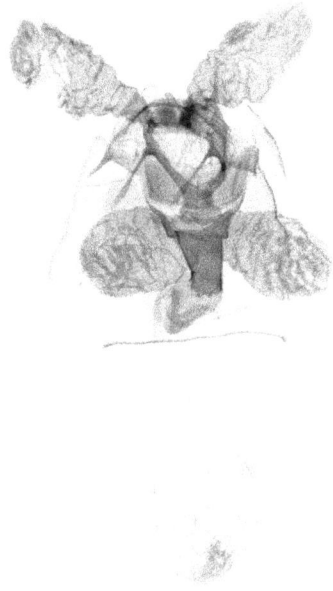

57

59

71

Plate 26

56. *Melisa croceipes* (AURIVILLIUS, 1892); DRC, „Ukaika"; coll. NHMW. - **57.** *Melisa diptera* (WALKER, 1854); DRC, "Stanley Falls"; 1900; coll. BMNH. - **59.** *Melisoides lobata* STRAND, 1912; DRC, Eala; 03.1935; coll. RMCA. - **71.** *Metarctia (Collocalia) tenebrosa* LE CERF, 1922; Tanzania, "W Usambara Mts, Mazumbai U.F.S."; 21.01.1985; coll. ISEA.

73

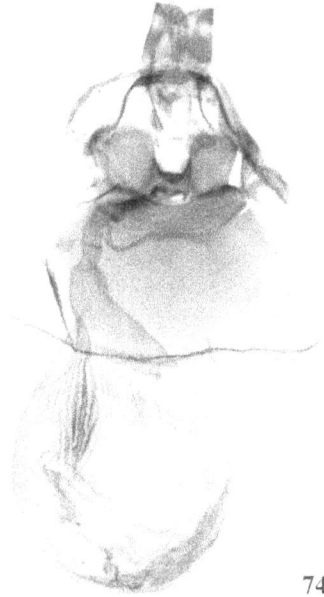

74

79

80

Plate 27

73. *Metarctia* (*Hebena*) *henrardi* KIRIAKOFF, 1953; paratype; coll. RMCA. - **74.** *Metarctia* (*Hebena*) *lateritia* (HERRICH-SCHAFFER, 1850-1858); Tanzania, "Mt. Meru, Momella 1600-1800m"; 10-19.02.1964; coll. ZSM. - **79.** *Metarctia* (*Metarctia*) *benitensis* HOLLAND, 1893; syntype; coll. CMNH. - **80.** *Metarctia* (*Metarctia*) *brunneipennis* HERING, 1932; holotype; coll. RMCA.

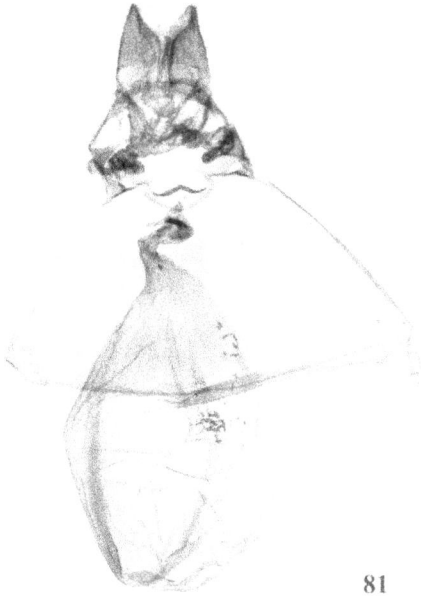

81

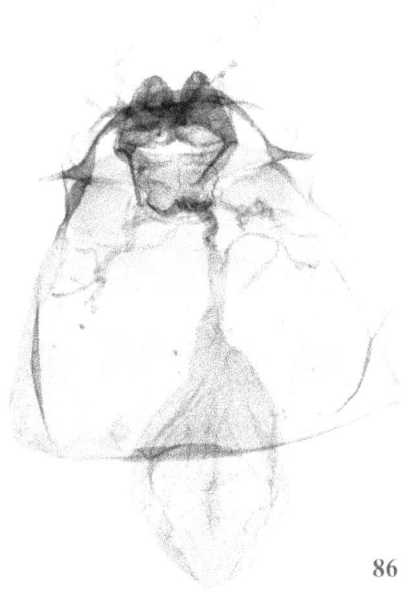

86

88

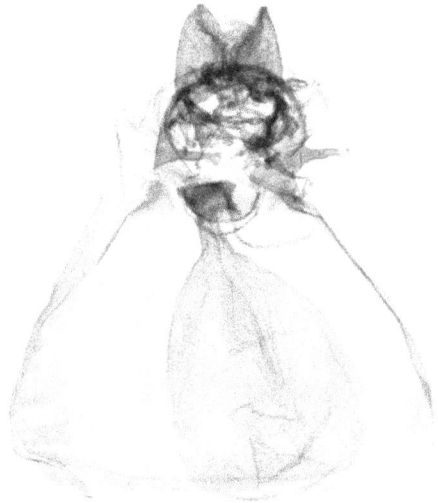

90

Plate 28

81._Metarctia(Metarctia)burra_(Schaus in Schaus & Clements, 1893); paratype of *chryseis* Kiriakoff, 1973; coll. ZSM. - **86.**_Metarctia(Metarctia)_
flaviciliata Hampson, 1907; Uganda, "Budongo"; 09.1932; coll. BMNH. - **88.** _Metarctia (Metarctia) flavivena_ Hampson, 1901; Burundi, „Kitega";
6.04.1964; coll. RMCA. - **90.** _Metarctia (Metarctia) fontainei_ Kiriakoff, 1953; DRC, "Sankuru, Katako-Kombe"; 3.03.1953; coll. RMCA.

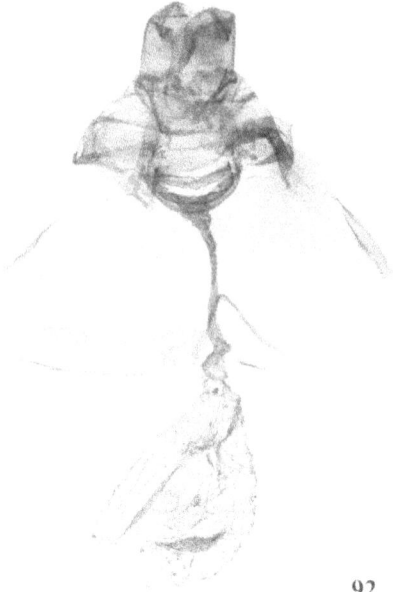

92

93

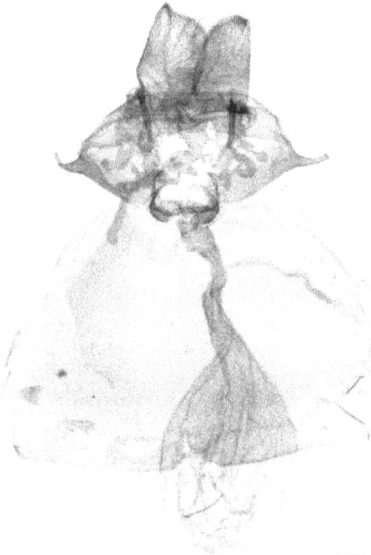

108

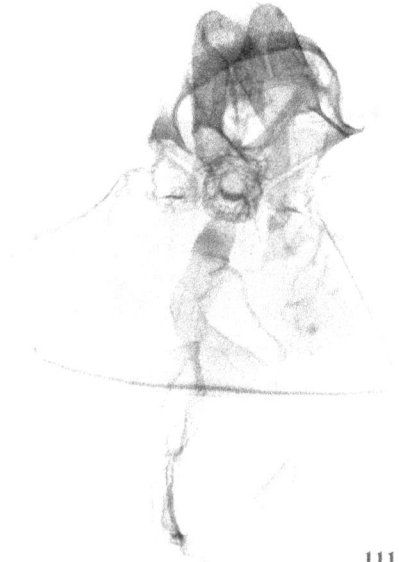

111

Plate 29

92. *Metarctia (Metarctia) fulvia* HAMPSON, 1901; lectotype of *neaera* FAWCETT, 1915; coll. BMNH. - **93.** *Metarctia (Metarctia) fusca* HAMPSON, 1901; Uganda; "Ruwenzori Range, Mahoma River, 6.700 ft"; 13-16.08.1952; coll. BMNH. - **108.** *Metarctia (Metarctia) nigritarsis* BERIO, 1943; paratype; coll. MCSNG. - **111.** *Metarctia (Metarctia) paremphares* HOLLAND, 1893; syntype; coll. CMNH.

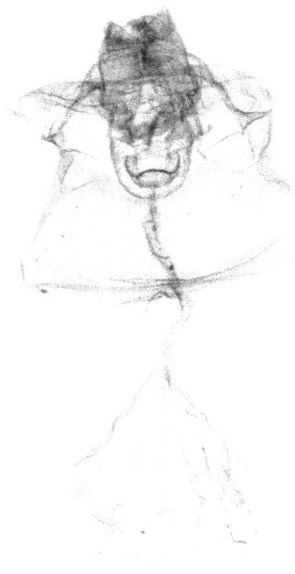

113

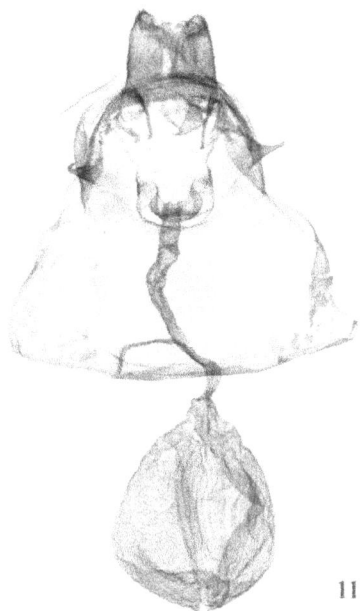

118

131

133

Plate 30

113. *Metarctia (Metarctia) phaeoptera* HAMPSON, 1909; holotype; coll. BMNH. - **118.** *Metarctia (Metarctia) rufescens* WALKER, 1855; RSA, "Elisabethville"; 12.04.1950; coll. RMCA. - **131.** *Metarctia (Metarctia) virgata* JOICEY & TALBOT, 1921; Uganda; "Ruwenzori Range, Mahoma River, 6.700 ft"; 13-16.08.1952; coll. BMNH. - **133.** *Metarctia (Metarhodia) epimela* (KIRIAKOFF, 1979); paratype; coll. ZSM.

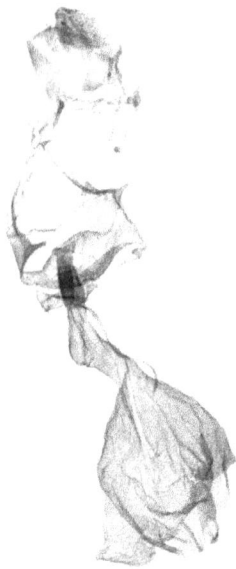

134

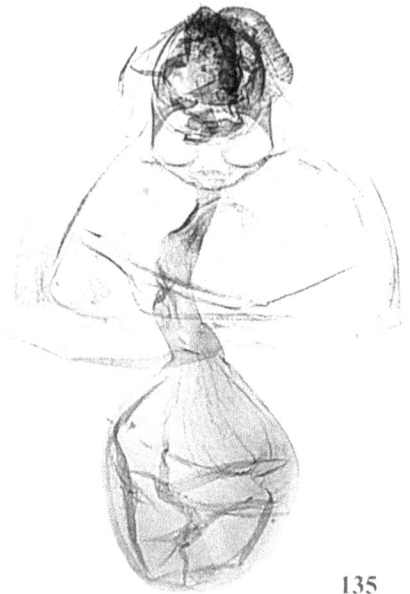

135

137

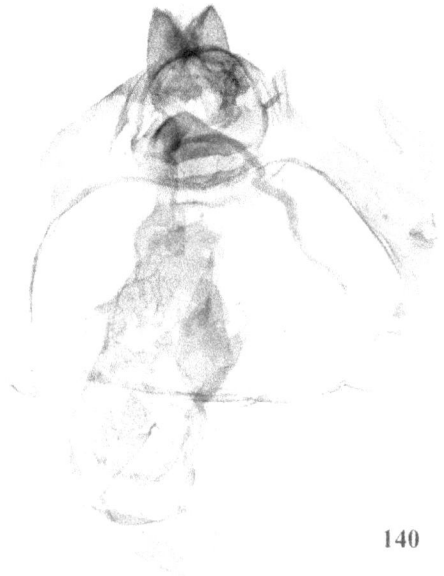

140

Plate 31

134. *Metarctia (Metarhodia) heinrichi* KIRIAKOFF, 1961; paratype; coll. ZSM. - **135.** *Metarctia (Metarhodia) heringi* KIRIAKOFF, 1957; paratype; coll. RMCA. - **137.** *Metarctia (Metarhodia) insignis* KIRIAKOFF, 1959; paratype; coll. RMCA. - **140.** *Metarctia (Metarhodia) rubribasa* BETHUNE-BAKER, 1911; Tanzania, "Matengo Hochland, wsw v. Songea, Ugano", 11-20.12.1935; coll. NHMW

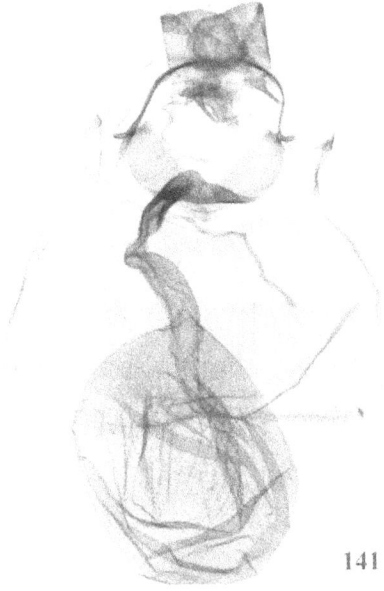

141

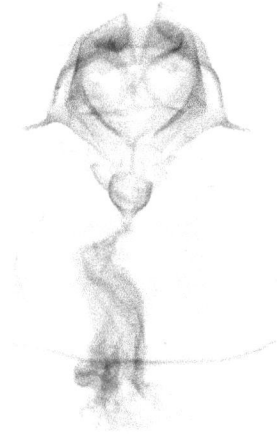

146

149

156

late 32

141. *Metarctia* (*Metarhodia*) *rubripuncta* HAMPSON, 1898; Cameroon, "Ari W. Kampf, D'dorf, Ekona"; 1.11.1936; coll. ZSM. - **146.** *Metarctia* (*Thyretarctia*) *haematica* HOLLAND, 1893; holotype of *haematosphages* HOLLAND, 1893; coll. CMNH. - **149.** *Metarctia* (*Thyretarctia*) *schoutedeni* KIRIA-FF, 1953; DRC, "Kibali-Ituri:Nioka"; 29.03.1954; coll. RMCA. - **156.** *Paramelisalophura* AURIVILLIUS, 1905; DRC, "Eala"; 01.1939; coll. RMCA.

158

159

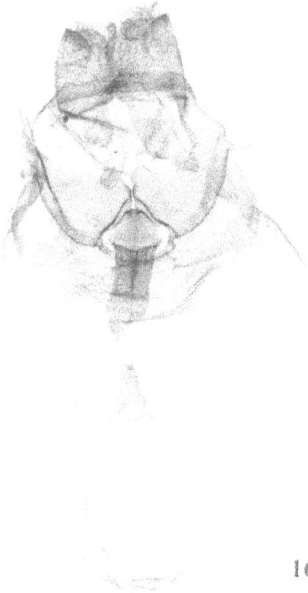

167

169

Plate 33

158. *Pseudmelisa chalybsa* HAMPSON, 1910; holotype; coll. BMNH. -**159.***Pseudmelisa rubrosignata* KIRIAKOFF, 1957; paratype; coll.BMNH.-1◆
Rhabdomarctia rubrilineata (BETHUNE-BAKER, 1911); Kenya, "WesternProv., KakamegaForestNR"; 16.04.2002; coll.ZMHB. -**169.***Rhipidarctia* (*El-sitia*) *forsteri* (KIRIAKOFF, 1953); Kenya, "WesternProv., Kakamega ForestN.R., prim. forest, 1600m, 0,21,34 N 34,51,39 E; 12.11.2001; coll.SMNS

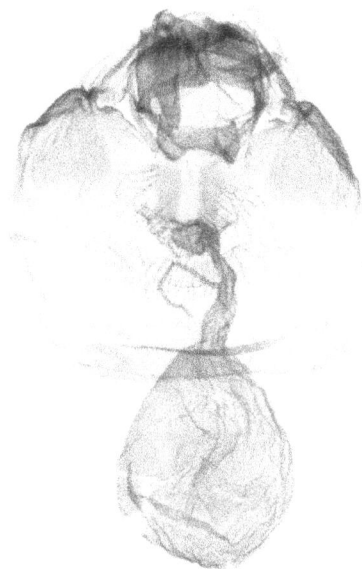

170

171

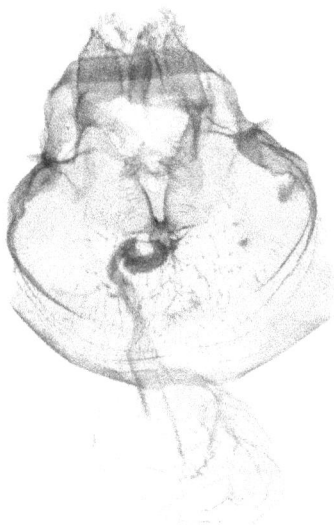

174

175

Plate 34

170. *Rhipidarctia (Elsitia) lutea* (HOLLAND, 1893); DRC, "Eala"; 10.1935; coll. RMCA. - **171.** *Rhipidarctia (Elsitia) pareclecta* (HOLLAND, 1893); Kenya, "Western Prov., Kakamega Forest N.R., prim. forest, 1600m, 0,21,34N 34,51,39E; 1.08.2002; coll. SMNS. - **174.** *Rhipidarctia (Hemirhipidia) postrosea* (ROTHSCHILD, 1913); Nigeria, "Anambra State, Nsukka F. Res."; 10.11.1982; coll. ISEA. - **175.** *Rhipidarctia (Rhipidarctia) aurora* KIRIAKOFF, 1957; DRC, Flandria [Boteka]; 08.1938; coll. RNHL.

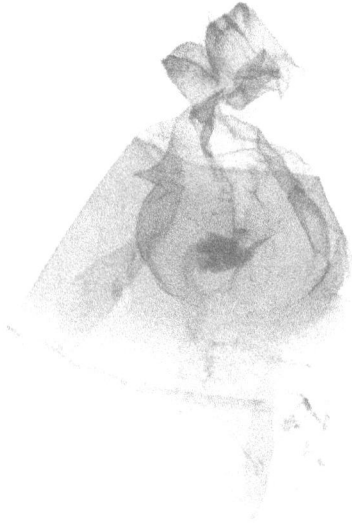

176

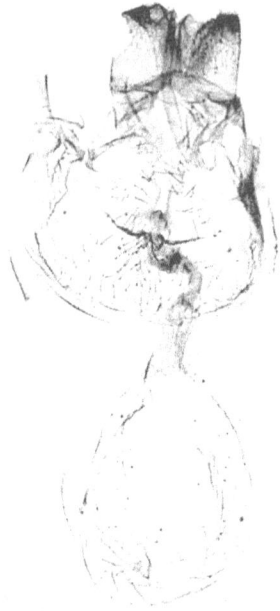

177

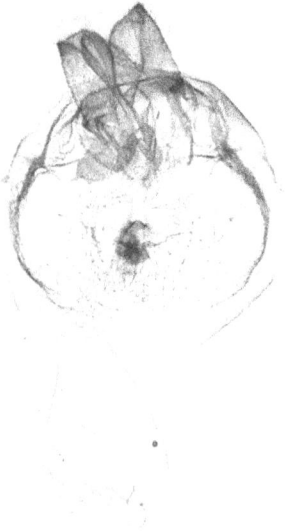

178

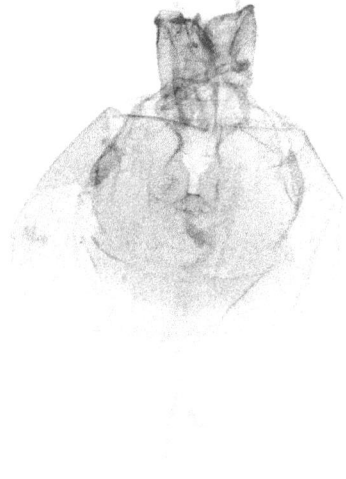

179

Plate 35

176. *Rhipidarctia (Rhipidarctia) conradti* (OBERTHUR, 1911); syntype; coll. BMNH. - **177.** *Rhipidarctia (Rhipidarctia) crameri* KIRIAKOFF, 1961; DRC, Uele, Paulis; 19.06.1956; coll. RMCA. - **178.** *Rhipidarctia (Rhipidarctia) flaviceps* (HAMPSON, 1898); DRC, Sankuru, Katako-Kombe; 9.03.1952; coll. RMCA. - **179.** *Rhipidarctia (Rhipidarctia) invaria* (WALKER, 1856); holotype; coll. BMNH.

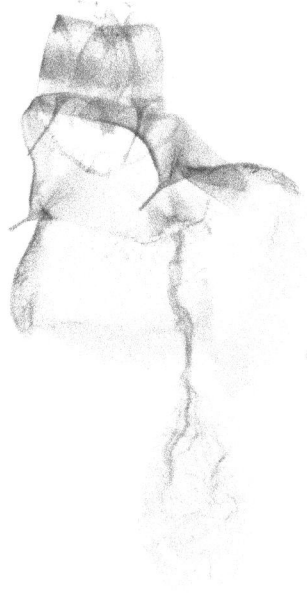

180

183

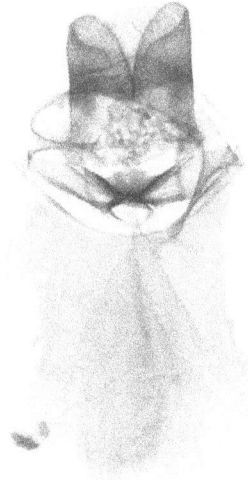

187

189

Plate 36

180. *Rhipidarctia (Rhipidarctia) miniata* KIRIAKOFF, 1957; holotype; coll. BMNH. - **183.** *Rhipidarctia rubrovitta* (AURIVILLIUS, 1904); syntype; coll. NHRS. - **187.** *Thyretes caffra* WALLENGREN, 1863; RSA, „Modderfontein, Transvaal"; 01.1921; coll. BMNH. - **189.** *Thyretes hippotes* (CRAMER, [1775-80]); RSA, "Cape Town"; 10.1861; coll. BMNH.

190

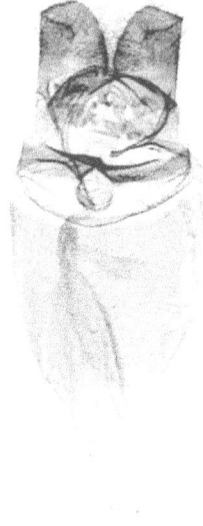

191

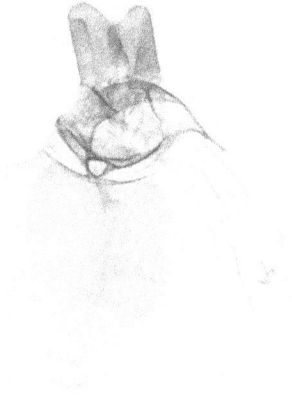

192

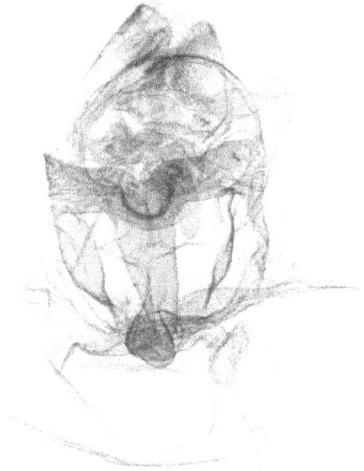

194

Plate 37

190. *Thyretes montana* BOISDUVAL, 1847; RSA, "Empangeni"; 3.02.1911; coll. TMSA. - **191.** *Thyretes monteiroi* BUTLER, 1876; holotype; coll. BMNH. - **192.** *Thyretes negus* OBERTHÜR, 1878; paralectotype of *misa* STRAND, 1911; coll. ZMHB. - **194.** *Pseudmelisa demiavis* KAYE, 1919; holotype; coll. BMNH.

INDEX OF SCIENTIFIC NAMES IN LEPIDOPTERA

A

abyssinibia KIRIAKOFF, 1957 23

abyssinibia STRAND, 1920 30

aegrota BERIO, 1939 23, 82

aethiops KIRIAKOFF, 1973 18, **58**, 59

affinis ROTHSCHILD, 1910 17, 38, **50**

alberici DUFRANE, 1945 (*Balacra*) 12, **44**

alberici DUFRANE, 1945 (*Pseudodiptera*) 32

albomaculata KIRIAKOFF, 1957 32

alticola AURIVILLIUS, 1925 19, **64**, 77, 83

Anace 11, 13, 19, 26, 28, 36

angolensis GAEDE, 1926 29

angolensis KIRIAKOFF, 1961 (*Balacra*) 14

angolensis KIRIAKOFF, 1957 (*Neophemula*) 25, **90**

Apisa **WALKER, 1855** 10, 11, 16, 26, 30, 31, 37-40

arabica WARNECKE, 1934 10, **37**

ashantica STRAND, 1916 30

atavistis HAMPSON, 1911 17

atrivenata KIRIAKOFF, 1956 19, **64**, 66, 77, 80, 81

aurantiifusca ROTHSCHILD, 1913 26

aurantiiventris KIRIAKOFF, 1953 32

aurivilliusi KIRIAKOFF, 1957 12

aurora KIRIAKOFF, 1957 28, **98**, 99, 100

Automolis **HÜBNER, [1819]** 11, 12, 18, 19, 21, 22, 23, 24, 34, 41, 42, 81

avellana KIRIAKOFF, 1957 20, **59**

B

Balacra **WALKER, 1856** 11, 12, 13, 14, 15, 17, 20, 30, 31, 34, 51, 101

Balacrella KIRIAKOFF, 1957 13, 34, 35

basilewskyi KIRIAKOFF, 1953 14, **48**, 50

batesi DRUCE, 1910 14, **49**

belga KIRIAKOFF, 1954 12, 42, 43

benitensis HOLLAND, 1893 19, 22, **65**, 74, 76

bergeri KIRIAKOFF, 1953 16

Bergeria **KIRIAKOFF, 1952** 15, 16, 51-53, 55

bicolora WALKER, 1856 11, **41**, 75, 81

bipartita KIRIAKOFF, 1973 27

bipuncta JOICEY & TALBOT, 1924 22

bitjeana BETHUNE-BAKER, 1927 17

bourgognei KIRIAKOFF, 1952 (*Apisa*) 10, 11

bourgognei KIRIAKOFF, 1952 (*Bergeria*) 15, **51**, 52, 55

brunnea GRÜNBERG, 1907 30

brunneipennis HERING, 1932 19, **65**

brunneoaurantiaca KIRIAKOFF, 1973 24, **87**

buettikeri WILTSHIRE, 1983 29, **102**, 104

bumba STRAND, 1918 13

burgessi KIRIAKOFF, 1957 16, 56

burra SCHAUS in SCHAUS & CLEMENTS, 1893 19, 22, 30, **65**, 66, 69, 80

burungae DEBAUCHE, 1942 19, **66**, 73, 70, 77

C

caeruleifascia WALKER, 1856 11, 12, 34, **43**

caffra WALLENGREN, 1863 29, **102**, 103, 104

Callobalacra **KIRIAKOFF, 1953** 12, 44, 45

Cameroonia PRZYBYŁOWICZ [gen. nov.] 15, 35, 53

cameruna HAMPSON, 1914 16, 54

cana HOLLAND, 1893 10

canescens WALKER, 1855 10, 30, 31, **37**, 40

capricornis KIRIAKOFF, 1957 22

carmel KIRIAKOFF, 1957 20, **66**

carnea HAMPSON, 1898 26, **93**, 94

Ceryx 32

chalybea ZERNY, 1912 26

chalybsa HAMPSON, 1910 26, 92, 93

chapini HOLLAND, 1920 24

christophi PRZYBYŁOWICZ, 2005 17, **58**

chryseis KIRIAKOFF, 1973 19, 65

cinctella KIRIAKOFF, 1953 27, **96**, 97

cinereocostata HOLLAND, 1893 11, 39, **40**

cinereoguttata STRAND, 1912 15, **53**, 54, 55

cinnamomea WALLENGREN, 1860 19, **62**, 63, 64

clypeatus KIRIAKOFF, 1965 33

Collartisa KIRIAKOFF, 1953 17

collartorum KIRIAKOFF, 1953 17

collocalia KIRIAKOFF, 1957 20, **59**, 60, 61, 62

Collocaliodes KIRIAKOFF, 1957 18, 58-62, 83

compsa JORDAN, 1904 13, **45**

Compsochromia KIRIAKOFF, 1953 13, 45

confederationis KIRIAKOFF, 1961 23, **83**, 84, 85, 86

congoensis KIRIAKOFF, 1957 25, **90**

congoensis ROTHSCHILD, 1910 14

congonis STRAND, 1916 30

conradti OBERTHÜR, 1911 12, 28, **98**, 99

contrasta BETHUNE-BAKER, 1911 20

cooremani KIRIAKOFF, 1953 29, **102**, 104

cornelia KIRIAKOFF, 1957 28

costalis KIRIAKOFF, 1973 27

crameri KIRIAKOFF, 1961 28, 97, 98, **99**, 100

crassa FELDER, 1874 11, 21, **41**

croceipes AURIVILLIUS, 1892 17, **56**, 57

crocina KIRIAKOFF, 1973 24, **87**

curriei DYAR, 1899 13

cyaneotincta HAMPSON, 1918 33

D

damalis HOLLAND, 1893 13

danieli KIRIAKOFF, 1955 28

daphaena HAMPSON, 1898 13, 34, **46**

Daphaenisca KIRIAKOFF, 1953 **13**, 34, 35, 46

debauchei KIRIAKOFF, 1953 18, **59**

Decimia WALKER, 1856 11

decora OBERTHÜR, 1911 14

demiavis KAYE, 1919 30, **105**

denisae DUFRANE, 1945 24

deriemaeckeri KIRIAKOFF, 1953 24, 86

Diakonoffia KIRIAKOFF, 1953 26

diaphana KIRIAKOFF, 1957 13, **45**

didyma KIRIAKOFF, 1957 24, **88**

diffusa DUFRANE, 1952 16

diptera WALKER, 1854 16, 17, 56, **57**

distincta KIRIAKOFF, 1953 14

diversa BETHUNE-BAKER, 1911 20, **66**, 82

dollmani HAMPSON, 1920 25, **91**

dracoena KIRIAKOFF, 1953 18, **60**, 61

dracuncula KIRIAKOFF, 1957 20

Dufraneella KIRIAKOFF, 1953 10, 38, 39

dufranei KIRIAKOFF, 1952 16, **55**, 56

dufranei KIRIAKOFF, 1965 33

E

efulensis HOLLAND, 1898 32, 33

ehrmanni HOLLAND, 1893 12

elegans AURIVILLIUS, 1892 13, 30, 46, **47**

elegantissima STRAND, 1912 30

Elsa KIRIAKOFF, 1953 27, 28, 36

Elsita KIRIAKOFF, 1954 28, 36

Elsitia PRZYBYŁOWICZ **[nom. nov.]** 27, 36, 95-97

Epibalacra KIRIAKOFF, 1957 14, 34

epimela KIRIAKOFF, 1979 23, **84**, 85, 86

Eressades BETHUNE-BAKER, 1911 29

erlangeri ROTHSCHILD, 1910 19, 23, 35, 82

erubescens HAMPSON, 1901 26, **93**

erubescens JOICEY & TALBOT, 1924 12

erubescens WALKER, 1864 28

Euchromia 16, 17

F

fario KIRIAKOFF, 1957 20, **67**, 75, 76, 79

fenestrata JORDAN, 1904 13

ferrigera DRUCE, 1910 16, **54**

flaviceps HAMPSON, 1898 28, 98, **99**, 100

flaviciliata HAMPSON, 1907 20, **67**, 80

flavicincta AURIVILLIUS, 1900 20, **67**, 72, 79

flavimacula WALKER, 1856 14, 30, 31, 34, 48, **49**, 50, 51

flavipunctata BETHUNE-BAKER, 1911 29

flavivena HAMPSON, 1901 19, 20, **68**

flavivena HAMPSON, 1902 20

fletcheri KIRIAKOFF, 1957 15, 51, 52

fletcheri KIRIAKOFF, 1958 21

flora KIRIAKOFF, 1957 20, 64, **68**, 77, 83

fontainei KIRIAKOFF, 1953 (*Balacra*) 14, **49**

fontainei KIRIAKOFF, 1953 (*Metarctia*) 20, 67, **69**, 80

fontainei KIRIAKOFF, 1959 10, **38**, 39, 69

forsteri KIRIAKOFF, 1953 27, **96**

forsteri KIRIAKOFF, 1955 20, **69**

fuliginosa KIRIAKOFF, 1953 18, 58, 59, **60**, 61, 62

fulvia HAMPSON, 1901 20, 65, 66, **69**, 70, 79

furva HAMPSON, 1911 14, **47**, 48

fusca HAMPSON, 1901 21, **70**

fuscorufescens STRAND, 1916 30

G

galla ROUGEOT, 1977 21, **70**

germana ROTHSCHILD, 1912 12

ghesquierei KIRIAKOFF, 1953 27

glagoessa HOLLAND, 1893 14

gloriosa JORDAN, 1904 12

grandis HOLLAND, 1893 17

grisescens DUFRANE, 1945 10, **38**

guillemei OBERTHÜR, 1911 12, **43**

H

haemalea HOLLAND, 1893 14, 49, **50**

haematica HOLLAND, 1893 24, **88**, 89

haematochrysia KIRIAKOFF, 1952 15, **51**, 52

haematoessa HOLLAND, 1893 16, 54, **56**

haematosphages HOLLAND, 1893 24

haematricha HAMPSON, 1905 21, **70**, 78

hampsoni KIRIAKOFF, 1953 33

hampsoni STRAND, 1916 30

hancocki JORDAN, 1936 17, **57**

Hebena WALKER, **1856** 18, 62-64, 83

hebenoides KIRIAKOFF, 1973 21, **71**

hecqi KIRIAKOFF, 1959 21

hector KIRIAKOFF, 1959 19, 65

heinrichi KIRIAKOFF, 1961 23, 83, **84**, 85, 86

helga KIRIAKOFF, 1957 22

Hemirhipidia KIRIAKOFF, 1955 26, 98

henrardi KIRIAKOFF, 1953 18, **62**, 64

heringi KIRIAKOFF, 1957 23, **80**

herona DRUCE, 1887 13, **46**, 47

Heronina KIRIAKOFF, 1955 **13**, 46

hewitti JANSE, 1945 11

Hexaneura WALLENGREN, 1860 17, 22

hildae KIRIAKOFF, 1961 6, 38, **39**

hippotes CRAMER, [1775-80] 29, **103**

Hippurarctia KIRIAKOFF, 1953 15, 16, 53, 54

homoerotica STRAND, 1916 30

hulstaertiana KIRIAKOFF, 1953 21, 69, **71**, 74

humphreyi ROTHSCHILD, 1912 14, **50**

hypomela KIRIAKOFF, 1956 23, 84, **85**

I

impura KIRIAKOFF, 1959 24, 86

incensa WALKER, 1864 11, **41**, 42

inconspicua HOLLAND, 1892 21, **71**

inconspicua HOLLAND, 1896 21

infausta KIRIAKOFF, 1957 25, **88**, 89

inflammata HAMPSON, 1914 12

insignis KIRIAKOFF, 1959 24, 84, **85**, 86

intermedia ROTHSCHILD, 1912 14

invaria WALKER, 1856 27, 28, 31, 36, **99**, 100

J

jacksoni KIRIAKOFF, 1956 20

jaensis BETHUNE-BAKER, 1927 12, **44**

jansei KIRIAKOFF, 1957 18, 59, **60**

johanna KIRIAKOFF, 1979 21, **72**

jordani KIRIAKOFF, 1957 24, **85**

jubdoensis KIRIAKOFF, 1955 21

judith KIRIAKOFF, 1959 16, 53, **54**

K

kamitugensis DUFRANE, 1945 (*Hippurarctia*) 16, 31

kamitugensis DUFRANE, 1945 (*Pseudothyretes*) 26, **93**

kamitugensis DUFRANE, 1945 (*Rhabdomarctia*) 27

katriona KIRIAKOFF, 1957 20

kelleni SNELLEN, 1886 19

kenyae KIRIAKOFF, 1957 23

kilimaensis KIRIAKOFF, 1973 18, **59**

kivensis DUFRANE, 1945 (*Balacra*) 30

kivensis DUFRANE, 1945 (*Pseudothyretes*) 26

kumasina STRAND, 1920 21, 70, **72**

L

Lamprobalacra KIRIAKOFF, 1953 13, 47, 48

lateritia HERRICH-SCHÄFFER, 1850-1858 18, 23, 30, 62, **63**, 68, 82

lateritiola STRAND, 1916 30

latipennis KIRIAKOFF, 1957 21

latophaga DUFRANE, 1945 30

laureola DRUCE, 1910 15

Lempkeella KIRIAKOFF, 1953 16, 51, 55

leroyi KIRIAKOFF, 1953 26, **71**

lindemannae KIRIAKOFF, 1961 21, **72**

lippensi KIRIAKOFF, 1960 10, 37

lobata STRAND, 1912 17, **57**

longimaculata STRAND, 1912 31

longipalpis HULSTAERT, 1923 21, 64, **73**, 81

lophura AURIVILLIUS, 1905 25, 26, **91**

lophuroides OBERTHÜR, 1911 26, **92**

luctuosa KIRIAKOFF, 1972 25, **89**

lugubris GAEDE, 1926 21, **73**, 74

lutea HOLLAND, 1893 27, **96**

M

maculifera WALLENGREN, 1860 22

magna HULSTAERT, 1923 12

major LE CERF, 1922 23, 82

manettii TURATI, 1924 11, **40**

margaretha KIRIAKOFF, 1957 18, 59

maria KIRIAKOFF, 1957 21, **73**

mariae DUFRANE, 1945 (*Apisa*) 30

mariae DUFRANE, 1945 (*Melisa*) 17

mariae DUFRANE, 1945 (*Pseudothyretes*) 26, **94**

Mecistorhabdia KIRIAKOFF, **1953** 16, 35, 56

Meganaclia AURIVILLIUS, **1892** 26, 32, 33

Megapisa AURIVILLIUS, 1904 12

melaena HAMPSON, 1905 13

melinos MABILLE, 1890 28

Melisa WALKER, **1854** 16, 17, 56, 57

Melisoides STRAND, **1912** 17, 57

Mesonaclia KIRIAKOFF, 1953 32

metaleuca HAMPSON, 1914 21, **74**, 80

Metamicroptera HULSTAERT, **1923** 17, 58, 92

Metapiconoma ROTHSCHILD, 1910 12

Metarctia WALKER, **1855** 7, 10, 11, 12, 13, 14, 15, 16, 17, 18, 19 - 31, 34, 35, 36, 58 - 89

Metaretia PAGENSTECHER, 1909 18

Metarhodia KIRIAKOFF, **1953** 23, 83 - 86

meteus STOLL, 1780-82 11, 34, **41**, 81

meteus WALKER, 1855 11

Microbergeria KIRIAKOFF, **1972** 25, 89

microcanescens BERIO, 1935 10

micromacula STRAND, 1920 31

Micrometaptera KIRIAKOFF, 1960 17

microsippia STRAND, 1912 32, 33

miniata KIRIAKOFF, 1957 28, 99, **100**

minor HAMPSON, 1914 32, 33

misa STRAND, 1911 29, 104

moira KIRIAKOFF, 1957 20

monotonia STRAND, 1912 14, 49, **50**, 51

montana BOISDUVAL, 1847 29, **103**

monteiroi BUTLER, 1876 29, **103**

montium KIRIAKOFF, 1957 19, **595**

morag KIRIAKOFF, 1957 21, **74**

morosa KIRIAKOFF, 1957 25, **89**

musiforme KAYE, 1918 32, 33

N

Naclia 32, 33

Nacliodes STRAND, **1912** 32, 33

neaera FAWCETT, 1915 20, 70

negus OBERTHÜR, 1878 29, 102, 103, **104**

negusi KIRIAKOFF, 1957 21, 67, **74**

Neobalacra KIRIAKOFF, 1952 17

Neophemula KIRIAKOFF, **1957** 25, 90

nigriceps Aurivillius, 1904 15, 35, **53**

nigricornis Debauche, 1942 24, **86**

nigripennis Aurivillius, 1904 12, 42, **43**

nigrita Kiriakoff, 1961 26, 89, **94**, 95

nigritarsis Berio, 1943 21, **75**

noctis Druce, 1910 22, **75**

Notharctia Kiriakoff, 1953 19

nyasae Kiriakoff, 1957 10

O

obliterata Grünberg, 1907 31

obscura Dufrane, 1945 31

occidentalis Kiriakoff, 1957 15, **52**

ochracea Walker, 1869 12

ochreogaster Joicey & Talbot, 1921 27

octava Kiriakoff, 1961 15, 51, **52**

Oenarctia Kiriakoff, 1953 19, 31, 35

olbrechtsi Kiriakoff, 1953 19, **61**

opobensis Strand, 1916 31

oreophila Kiriakoff, 1963 14

orientalis Kiriakoff, 1956 22

ornata Kiriakoff, 1959 15, 51, **52**

overlaeti Kiriakoff, 1953 16

Owambarctia Kiriakoff, **1957** 25, 87, 90

owamboensis Kiriakoff, 1957 25, **90**

P

Pachyceryx Kiriakoff, 1957 32, 33

pallata Plötz, 1880 10

pallens Bethune-Baker, 1911 22, 67, 74, **75**, 76, 79

pallida Hampson, 1901 11, 21, 34, **42**, 72, 80

pallidicosta Hulstaert, 1923 20

pallidipes Aurivillius, 1925 30

pamela Kiriakoff, 1957 21

paniscus Kiriakoff, 1957 19

panyamana Strand, 1920 20

paradoxa Hering, 1932 17

paradoxa Romieux, 1934 17

Paramelisa Aurivillius, **1905** 17, 25, 26, 57, 91 - 92

Parapisa Kiriakoff, **1952** 10, 40

pareclecta Holland, 1893 27, 36, **97**

paremphares Holland, 1893 22, 65, 74, **76**

paulis Kiriakoff, 1961 22, **76**

pavlitzkae Kiriakoff, 1961 18, 59, 60, **61**

perpusilla Walker, 1856 26, **94**, 95

perversa Strand, 1916 31

phaeoptera Hampson, 1909 22, **77**, 78, 81

phasma Butler, 1897 29

Pinheya Kiriakoff, 1973 24

Pinheyata Nye in Watson, Fletcher & Nye, **1980** 24, 35, 87

pinheyi Kiriakoff, 1956 25, 89

Plegapteryx 29

postfuscescens Strand, 1916 31

postrosea Rothschild, 1913 28, **98**

preussi Aurivillius, 1904 14, 15, 30, 31, 34, 43, 49

priscilla Kiriakoff, 1957 22, **77**

Pseudapiconoma Aurivillius, **1881** 12, 13, 14, 15, 16, 25, 30, 31, 34, 48 - 51

Pseudmelisa Hampson, **1910** 26, 30, 60, 92, 105

Pseudodiptera Kaye, **1918** 32, 33

Pseudomelisa Zerny, 1912 26, 30

Pseudothyretes Dufrane, **1945** 26, 89, 93

Psychotoe 10

Pterophaea KIRIAKOFF, 1953 19

pulchra AURIVILLIUS, 1892 23, 14, **49**

pulverea HAMPSON, 1907 22, **77**

pumila HAMPSON, 1909 22, **78**, 79, 80

punctata DUFRANE, 1945 31

punctulata KIRIAKOFF, 1963 27, 96

pusillima STRAND, 1912 31

Q

quadrisignatula STRAND, 1912 31

quinta KIRIAKOFF, 1973 24, **87**

R

rattrayi ROTHSCHILD, 1910 12, 43, **44**

rendalli ROTHSCHILD, 1910 10, 38, **39**

***Rhabdomarctia* KIRIAKOFF, 1953** 27, 95

***Rhipidarctia* KIRIAKOFF, 1953** 7, 27, 28, 29, 36, 95 - 101

rhodites KIRIAKOFF, 1957 19

rhodosoma KIRIAKOFF, 1957 29, 98, 99, **100**

rhodospila KIRIAKOFF, 1957 28

robusta KIRIAKOFF, 1973 22, 71, **77**

rosacea BETHUNE-BAKER, 1911 27

rosea AURIVILLIUS, 1905 31

rothschildi LE CERF, 1922 20

rotundata HULSTAERT, 1923 17, **58**

ruandae KIRIAKOFF, 1961 27

rubicundula STRAND, 1912 26, 30, **95**

rubra WALKER, 1856 19, 62, **63**

rubribasa BETHUNE-BAKER, 1911 24, 84, 85, **86**

rubricincta HOLLAND, 1893 14, 30, 47, **47**

rubricosta KIRIAKOFF, 1957 31

rubricosta TALBOT, 1929 24, 31

rubrilineata BETHUNE-BAKER, 1911 27, **95**

rubripuncta HAMPSON, 1898 23, 24, 30, 31, 84, 85, **86**

rubrosignata KIRIAKOFF, 1957 26, **92**

rubrostriata AURIVILLIUS, 1898 12, 44, **45**

rubrosuffusa KIRIAKOFF, 1953 27, 28, 36

rubrovitta AURIVILLIUS, 1904 14, 29, 97, **101**

rufescens WALKER, 1855 17, 19, 22, 30, 31, 35, 64, 70, 72, **78**, 87

S

saalfeldi KIRIAKOFF, 1960 22, **79**

salmonea KIRIAKOFF, 1957 22, 67, 74, **79**

sarcosoma HAMPSON, 1901 22, 67, 69, **79**

saturata KIRIAKOFF, 1957 27, 96, **97**

schoutedeni KIRIAKOFF, 1952 15, **52**

schoutedeni KIRIAKOFF, 1953 25, 35, 87, 88, **89**

separata STRAND, 1916 31

septentrionalis KIRIAKOFF, 1957 16

seydeliana KIRIAKOFF, 1953 18, 60, **61**

sheljuzhkoi KIRIAKOFF, 1961 22, **80**

signivenis HERING, 1937 30, 103, **104**

silacea PLÖTZ, 1880 29, **101**

similis HULSTAERT, 1923 12

similis KIRIAKOFF, 1953 27

simplex AURIVILLIUS, 1925 14

simplicior KIRIAKOFF, 1957 14

sippia PLÖTZ, 1880 32, 33

speculigera GRÜNBERG, 1907 15, 31

Sphinx 11, 29, 34

stigmatica GRÜNBERG, 1907 13

strenua KIRIAKOFF, 1957 27, 96

subargentea JOICEY & TALBOT, 1921 11, **40**

subcanescens ROTHSCHILD, 1910 10, 38, **39**

subincarnata KIRIAKOFF, 1954 19, 62, **64**

subminiata KIRIAKOFF, 1959 28, 96, **97**

subnigra KIRIAKOFF, 1958 19

subpallens KIRIAKOFF, 1956 21, 78, 79, **80**

subpumila KIRIAKOFF, 1957 20

subrosea KIRIAKOFF, 1957 11

sudanica KIRIAKOFF, 1973 19, 66

syntomia PLÖTZ, 1880 29, **101**

Syntomis 32, 33

T

Takwa KIRIAKOFF, 1957 28, 29, 36, 101

tamsi KIRIAKOFF, 1952 15, **53**

tamsi KIRIAKOFF, 1957 10

taymansi ROTHSCHILD, 1910 16, 31, **54**

tenebrosa LE CERF, 1922 18, 60, 61, **62**

tenera KIRIAKOFF, 1973 23, **80**

testacea AURIVILLIUS, 1881 14, 31, 34

Thyretarctia STRAND, **1912** 24, 35, 70, 87 - 89

Thyretes BOISDUVAL, **1847** 7, 28, 29, 30, 31, 102 - 104

Thyrogonia HAMPSON, **1898** 32, 33

titan TALBOT, 1929 19

transvaalica KIRIAKOFF, 1973 23, 81

trichaetiformis ZERNY, 1912 33

tricolor ROUGEOT, 1977 22

tricolorana WICHGRAF, 1922 23, 81

Tritonaclia 26

U

ugandae ROTHSCHILD, 1910 14

umbra DRUCE, 1910 15

umbretta KIRIAKOFF, 1963 19

unicolor KIRIAKOFF, 1957 27, 96

unicolor OBERTHÜR, 1880 23, 81, 82

uniformis BETHUNE-BAKER, 1911 23, **82**

unipuncta KIRIAKOFF, 1973 25, **90**

upembae KIRIAKOFF, 1954 23, **82**

usta DEBAUCHE, 1942 21

V

vanoyei KIRIAKOFF, 1952 16, **55**

venosa WALKER, 1856 18

venustissima KIRIAKOFF, 1961 23, **82**

vicina KIRIAKOFF, 1953 15, 16

virgata JOICEY & TALBOT, 1921 23, 77, **83**

vitreata ROTHSCHILD, 1910 13

vitreigutta HULSTAERT, 1923 15

vitrina OBERTHÜR, 1909 25, **90**

W

waelbroecki DEBAUCHE, 1938 27

wittei DEBAUCHE, 1942 23

X

xanthippa KIRIAKOFF, 1956 22

xenops KIRIAKOFF, 1957 28, 29, 36, **100**

Z

Zagaris WALKER, 1855 11

zegina STRAND, 1920 20, 68

www.ingramcontent.com/pod-product-compliance
Lightning Source LLC
Chambersburg PA
CBHW070727220326
41598CB00024BA/3327